Primary **Maths** for **Scotland**

2nd Level Maths
Textbook 2A

Series Editor: Craig Lowther

Authors: Antoinette Irwin, Carol Lyon,
Kirsten Mackay, Felicity Martin, Scott Morrow

Contents

1 Estimation and rounding — 2

1.1 Rounding whole numbers to the nearest 10 — 2

1.2 Rounding whole numbers to the nearest 100 — 4

1.3 Rounding decimal fractions to the nearest whole number — 6

1.4 Estimating the answer using rounding — 8

2 Number – order and place value — 10

2.1 Reading and writing whole numbers — 10

2.2 Representing and describing whole numbers — 12

2.3 Place value partitioning of numbers with up to five digits — 14

2.4 Comparing and ordering numbers in the range 0–10000 — 18

2.5 Reading and writing decimal fractions — 20

2.6 Representing and describing decimal fractions — 22

2.7 Comparing and ordering decimal fractions — 24

2.8 Recognising the context for negative numbers — 27

3 Number – addition and subtraction — 30

3.1 Mental addition and subtraction — 30

3.2 Adding and subtracting 1, 10, 100 and 1000 — 32

3.3 Adding and subtracting multiples of 100 — 34

3.4 Adding and subtracting by making 10s or 100s — 36

3.5 Adding and subtracting multiples of 1000 — 38

3.6 Adding and subtracting multiples of 10 and 100 — 40

3.7 Solving word problems — 42

3.8 Representing word problems in different ways — 44

3.9 Using non-standard place value partitioning — 46

3.10 Adding using a semi-formal written method — 48

3.11 Adding three-digit numbers using standard algorithms — 50

3.12 Representing and solving word problems — 52

4 Number – multiplication and division   **54**

4.1 Recalling multiplication and division facts for 2, 5 and 10   **54**

4.2 Recalling multiplication and division facts for 3   **56**

4.3 Solving multiplication problems   **58**

4.4 Multiplying whole numbers by 10 and 100   **60**

4.5 Dividing whole numbers by 10 and 100   **62**

4.6 Solving multiplication problems by partitioning   **64**

4.7 Solving multiplication and division problems   **66**

4.8 Recalling multiplication and division facts for 4   **68**

4.9 Solving division problems   **70**

4.10 Multiplying by changing the order of factors   **72**

4.11 Solving multiplication problems   **74**

4.12 Using known facts and halving to solve division problems   **76**

5 Multiples, factors and primes   **78**

5.1 Recognising multiples and factors   **78**

6 Fractions, decimal fractions and percentages   **80**

6.1 Comparing fractions   **80**

6.2 Identifying equivalent fractions pictorially   **83**

6.3 Identifying and creating equivalent fractions   **86**

6.4 Simplifying fractions   **88**

6.5 Writing decimal equivalents to tenths   **91**

6.6 Comparing numbers with one decimal place   **94**

6.7 Calculating a simple fraction of a value   **97**

7 Money   **100**

7.1 Writing amounts using decimal notation   **100**

7.2 Budgeting   **102**

7.3 Saving money   **105**

7.4 Profit and loss   **108**

8 Time   **110**

8.1 Telling the time to the minute – 12-hour clock   **110**

8.2 Converting between 12-h and 24-h time   **114**

8.3 Converting minute intervals to fractions of an hour   **116**

8.4 Calculating time intervals or durations   **118**

8.5 Speed, time and distance calculations   **120**

9 Measurement   **122**

9.1 Using familiar objects to estimate length, mass, area and capacity   **122**

9.2 Estimating and measuring length   **126**

9.3 Estimating and measuring mass   **129**

9.4 Converting units of length   **132**

9.5 Calculating the perimeter of simple shapes   **135**

9.6 Finding the area of regular shapes in square cm/m   **138**

9.7 Finding the volume of cubes and cuboids by counting cubes   **141**

9.8 Estimating and measuring capacity   **144**

10 Mathematics, its impact on the world, past, present and future   **146**

10.1 Mathematical inventions and different number systems   **146**

11 Patterns and relationships   **148**

11.1 Exploring and extending number sequences   **148**

12 Expressions and equations   **150**

12.1 Solving simple equations using known number facts   **150**

13 2D shapes and 3D objects   **152**

13.1 Drawing 2D shapes   **152**

13.2 Naming and sorting 2D shapes   **154**

13.3 Drawing 2D shapes and 3D objects   **157**

13.4 Describing and sorting prisms and pyramids   **160**

14 Angles, symmetry and transformation   **162**

14.1 Identifying angles   **162**

14.2 Using an eight-point compass   **164**

14.3 Plotting points using coordinates   **168**

14.4 Lines of symmetry   **170**

14.5 Creating designs with lines of symmetry   **173**

14.6 Measuring angles up to 180°   **176**

14.7 Understanding scale   **179**

15 Data handling and analysis   **182**

15.1 Reading and interpreting information   **182**

15.2 Organising and displaying data   **187**

15.3 Reading and interpreting pie charts   **190**

15.4 Collecting data   **192**

16 Ideas of chance and uncertainty   **196**

16.1 Predicting and explaining simple chance situations   **196**

1 Estimation and rounding

1.1 Rounding whole numbers to the nearest 10

We are learning to round numbers to the nearest ten.

Before we start

What is this number? **5492**
What does the 4 represent?
What does the 9 represent?
How many thousands are there in this number?
Write the number that would be 1000 more.

Rounding a number to the nearest ten means finding the ten that it is closest to.

Let's learn

The number 47 is between 40 and 50, but it is closer to the number 50 so we round 47 up to 50.

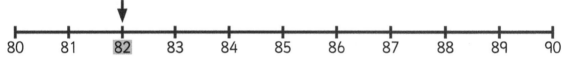

The number 82 is between 80 and 90, but it is closer to the number 80 so we round down to 80.

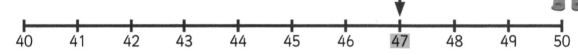

The number 458 is between 450 and 460, but it is closer to the number 460 so we round up to 460.

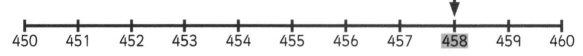

For numbers ending in 5 or higher we usually round up.

Let's practise

1) Use the number lines below to help you decide whether to round up or down by finding the nearest ten.

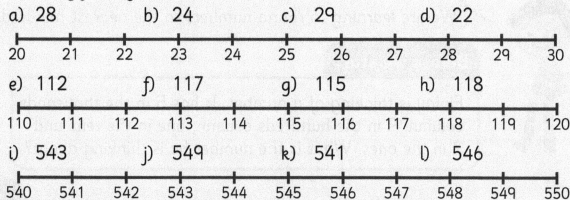

a) 28   b) 24   c) 29   d) 22

```
├────┼────┼────┼────┼────┼────┼────┼────┼────┼────┤
20   21   22   23   24   25   26   27   28   29   30
```

e) 112   f) 117   g) 115   h) 118

```
├────┼────┼────┼────┼────┼────┼────┼────┼────┼────┤
110  111  112  113  114  115  116  117  118  119  120
```

i) 543   j) 549   k) 541   l) 546

```
├────┼────┼────┼────┼────┼────┼────┼────┼────┼────┤
540  541  542  543  544  545  546  547  548  549  550
```

2) Are these number sentences true or false?

a) 23 rounded to the nearest ten is 20.

b) 85 rounded to the nearest ten is 80.

c) 154 rounded to the nearest ten is 160.

d) 371 rounded to the nearest ten is 360.

e) 1978 rounded to the nearest ten is 1980.

3) It is 356 km to the airport.

a) Which digit in the number do you look at to decide which ten 356 is closer to?

b) What is 356 rounded to the nearest ten? Explain your answer.

c) Write five more numbers that would round to the same answer as part b).

⭐ CHALLENGE!

In pairs, play the following game: nearest

1) Write down a secret three-digit number (for example, 433).

2) Round the number to the nearest 10 and tell your partner what the new number is (in this example it would be 430).

3) Your partner then has three attempts to try and guess the secret number. They get three points if they guess it first time, two points if they guess it second time and one point if they guess it third time.

4) Play the game several times and swap roles each time. To make things really tough, try rounding the number to the nearest 100!

1.2 Rounding whole numbers to the nearest 100

We are learning to round numbers to the nearest hundred.

Before we start

Finlay is thinking of a number. It has 5 in the thousands column, 9 in the hundreds column, one in the tens and 2 in the ones. What is the number he is thinking about?

Rounding a number to the nearest hundred means finding the hundred that it is closest to.

Let's learn

The number 178 is between 100 and 200, but it is closer to the number 200 so we round 178 **up** to 200.

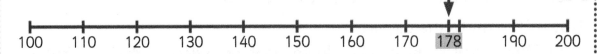

| 100 | 110 | 120 | 130 | 140 | 150 | 160 | 170 | 178 | 190 | 200 |

The number 412 is between 400 and 500, but it is closer to the number 400 so we round **down** to 400.

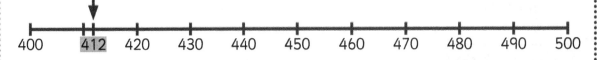

| 400 | 412 | 420 | 430 | 440 | 450 | 460 | 470 | 480 | 490 | 500 |

The number 3758 is between 3700 and 3800, but it is closer to the number 3800 so we round **up** to 3800.

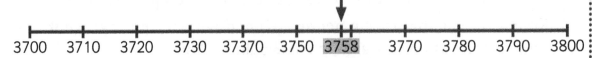

| 3700 | 3710 | 3720 | 3730 | 37370 | 3750 | 3758 | 3770 | 3780 | 3790 | 3800 |

For numbers ending in 50 or higher we usually round **up**.

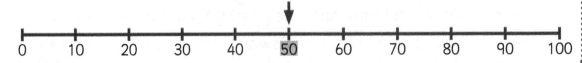

| 0 | 10 | 20 | 30 | 40 | 50 | 60 | 70 | 80 | 90 | 100 |

Let's practise

1) Write the two multiples of 100 on each side of the target number and state whether the number is rounded up or down.

a) 457 lies between 400 and 500   rounded up

b) 932 lies between ☐ and ☐ ☐

c) 643 lies between ☐ and ☐ ☐

d) 850 lies between ☐ and ☐ ☐

2) Are these number sentences true or false?
a) 236 rounded to the nearest 100 is 200 (True) / False
b) 754 rounded to the nearest 100 is 700 True / False
c) 815 rounded to the nearest 100 is 800 True / False
d) 278 rounded to the nearest 100 is 300 True / False
e) 547 rounded to the nearest 100 is 600 True / False

3) A theme park recorded the number of visitors each month. Copy and complete the table, rounding each number to the nearest 100.

Number of visitors	May	June	July	August	September
Actual numbers	793	996	2097	1071	617
Rounded to nearest hundred					

CHALLENGE!

Try these rounding puzzles for four-digit numbers. Start from the rounded figure and work in reverse to find the range of possible numbers.

a) Rounded to the nearest 10, the number becomes 1350. Rounded to the nearest 100, the number becomes 1300.

b) Rounded to the nearest 100, the number is 1200. Rounded to the nearest 10, the number is 1190. The number is a multiple of 5.

c) Create your own rounding puzzle for a partner to solve.

1.3 Rounding decimal fractions to the nearest whole number

We are learning to round decimal fractions to the nearest whole number.

Before we start

Write the decimal fraction that this diagram represents.
What number would be one tenth more?
One tenth less?

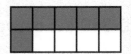

We can round decimal fractions up or down to the nearest whole number.

Let's learn

When we round decimal fractions to the nearest whole number we look at the first digit after the decimal point.

If this digit is 5 or more we usually round **up** to the nearest whole number.

If this digit is less than 5 we usually round **down** to the nearest whole number.

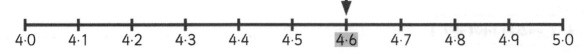

4·6 is between 4 and 5, but is closer to 5 so we round up to 5.

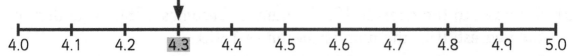

4.3 is between 4 and 5, but is closer to 4 so we round down to 4.

Let's practise

1) a) Write the numbers in the correct positions on the number line.

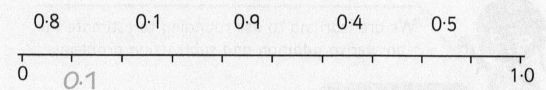

0·8   0·1   0·9   0·4   0·5

0   0·1   1·0

   b) Round each number from part a) to the nearest whole number.

2) Round each of these to the nearest whole number.

   a) 5.2   b) 6.5   c) 4.7   d) 6.4   e) 4.5

3) These are the average monthly warmest and coldest temperatures for Edinburgh for one year.

Round each of these temperatures to the nearest whole number.

January: 6.6 °C; 0.5 °C   July: 19.3 °C; 10.6 °C
February: 7.1 °C; 0.6 °C   August: 19.1 °C; 10.5 °C
March: 9.1 °C; 1.9 °C   September: 16.4 °C; 8.5 °C
April: 11.3 °C; 3.3 °C   October: 13.0 °C; 5.8 °C
May: 14.4 °C; 5.8 °C   November: 9.2 °C; 2.5 °C
June: 17.3 °C; 8.7 °C   December: 7.2 °C; 1.1 °C

CHALLENGE!

Find out the average temperatures to one decimal place for a city in a different part of the world. Round the temperatures to the nearest whole number.

Make a bar chart to show this information, and compare it to the temperatures for Edinburgh.

Does rounding to the nearest whole number make it easier to make a comparison? Explain your answer.

1 Estimation and rounding

1.4 Estimating the answer using rounding

We are learning to use rounding to estimate the answer to addition and subtraction problems.

Before we start

Nuria rounds the number 1547 to the nearest 10 and gets 1540. Is she correct? Explain.

Rounding makes it easier to estimate answers to problems.

Let's learn

We can round numbers to the nearest ten or hundred to help us get a quick estimate to addition and subtraction problems. This makes it easier to spot mistakes and check if our answer is reasonable.

Example

Amman takes everyone out to dinner. The cost of each meal is £18, £22, £27 and £19. Drinks cost £12. To estimate roughly how much the total will be, he rounds each amount to the nearest 10.

Cost of each meal (£)	18	22	27	19	12	
Rounded to the nearest £10	20	20	30	20	10	Estimate total = £100

He estimates he will need £100 to pay the bill. When he adds up the original prices it comes to £98 – his estimate was very close!

Let's practise

1) Estimate the answers by rounding to the nearest 10.

   a) 27 + 57 =   b) 83 + 58 =   c) 25 + 87 =

2) Isla and Finlay are organising a party.

Isla

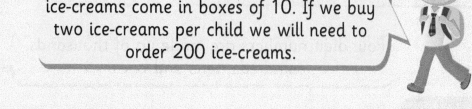

There are 25 family members and 211 friends coming over. Altogether there should be 461 people at the party. Oh! That can't be right!

Finlay

There will be 96 children at the party. The ice-creams come in boxes of 10. If we buy two ice-creams per child we will need to order 200 ice-creams.

   a) How did Isla use rounding to work out what was wrong with her first estimate? What should the answer be?

   b) How did Finlay work out the number of ice-creams to order?

⭐ **CHALLENGE!**

For the party, Isla and Finlay decorated a rectangular room.
- They put 332 flowers along the length and 421 flowers along the width.
- They also put 269 balloons along the length and 328 balloons along the width.
- They calculated they had used 1506 flowers and 1194 balloons.

Are these figures correct? Using your skills of rounding, produce estimates to check the reasonableness of the calculations.

1.4 Estimating the answer using rounding   9

Number – order and place value

2.1 Reading and writing whole numbers

We are learning to read and write four-digit numbers.

Before we start

Finlay thinks 7005 is seven hundred and five. Explain why Finlay is incorrect. Write the number seven hundred and five correctly. What is the value of zero?

Four-digit numbers are made up of thousands, hundreds, tens and ones.

Let's learn

We can write four-digit numbers using numerals and words. For example, 3564 can also be written as three thousand, five hundred and sixty-four.

Thousands		Ones		
	O	H	T	O
	3	5	6	4

When reading and writing four-digit numbers we must remember to use the word 'and' before the tens and ones. For example, 2461 is two thousand, four hundred and sixty-one.

Thousands		Ones		
	O	H	T	O
	2	4	6	1

Zero is a place holder in some four-digit numbers. For example, four thousand and two is written as 4002. The zeros keep the four thousands and two ones in place.

Thousands		Ones		
	O	H	T	O
	4	0	0	2

Let's practise

1) Write the number shown on each place value house in words.

a)

Thousands		Ones		
	O	H	T	O
	3	1	3	8

b)

Thousands		Ones		
	O	H	T	O
	6	7	2	4

c)

Thousands		Ones		
	O	H	T	O
	7	0	0	2

d)

Thousands		Ones		
	O	H	T	O
	5	5	0	0

2) Write these numbers in numerals and describe the position of the place holder. For example:

One thousand three hundred and four

1304 The place holder is in the **tens** position.

a) Two thousand and seventy-eight
b) Five thousand, three hundred and two
c) Six thousand, one hundred and sixty
d) Nine thousand and ninety-nine

3) Use these numerals to make as many four-digit numbers as you can.

Write your numbers in numerals and words.

7 0 4 3

⭐ **CHALLENGE!** ....

Amman called out this number and asked Nuria and Isla to write it in words.

1217

Nuria wrote one thousand, two hundred seventeen. Explain why she is incorrect.

Isla wrote one thousand, two hundred and seventy. Explain why Isla is incorrect.

Write what Nuria and Isla should have written.

2.2 Representing and describing whole numbers

We are learning to build and describe four-digit numbers.

Before we start

Represent the number 216 in three different ways. You may use concrete materials, drawings or symbols. Your representations should clearly show each digit's value.

The position of each digit tells us its value.

Let's learn

From left to right four-digit numbers have values of thousands (1000s), hundreds (100s), tens (10s) and ones (1s).

The same digit can have different values. For example, in the number **2351** the **5 is worth five tens (50)** but in the number **5132** the **5 is worth five thousands (5000)**.

We can use base 10 blocks or place value counters to show the values of the digits in a four-digit number. For example, 3113 can be made with three thousands, one hundred, one ten and three ones.

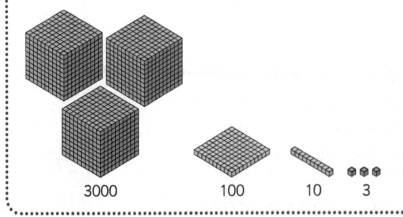

| 3000 | 100 | 10 | 3 |

Let's practise

1) Write the number represented by each set of base 10 blocks in numerals and in words.

a)
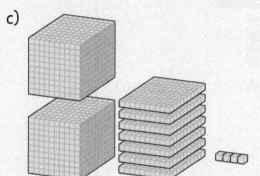

b)
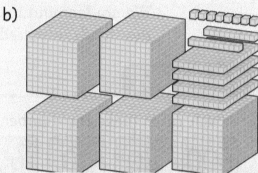

c)
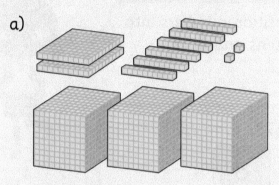

d)
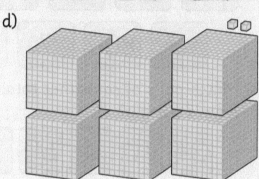

2) In the number 2605 the value of the 2 is two thousand or 2000. Write the value of the 2 in these numbers in both words and numerals.

a) 8283   b) **2**765   c) 15**2**3

d) 954**2**   e) 1**2**05   f) 400**2**

3) Use base 10 blocks or place value counters to build some four-digit numbers of your choice. Ask a partner to write your numbers in numerals and in words. Are they correct?

★ **CHALLENGE!** ....

Nuria's mystery four-digit number has a 6 in the ones place, a 3 in the thousands place, a placeholder in the tens place and a 5 in the hundreds place. Write Nuria's mystery number.

Nuria then took away a hundred and a thousand and added a ten. What number does she have now?

2.3 Place value partitioning of numbers with up to five digits

We are learning to partition numbers into thousands, hundreds, tens and ones.

Before we start

a) Which cards make the number 867?

| 600 | 80 | 70 | 60 | 8 |

| 6 | 800 | 7 | 700 |

b) Match pairs of cards that total 54. Which card doesn't have a partner?

| 34 | 10 | 4 | 50 | 5 |

| 30 | 44 | 40 | 14 | 24 | 20 |

We can break numbers up in different ways to make them easier to work with. This is called **partitioning**.

Let's learn

Let's partition 7345.

7 is in the thousands place. Its value is seven thousand, or 7000.

Thousands		Ones		
O	H	T	O
7	3	4	5

3 is in the hundreds place. Its value is three hundred, or 300.

4 is in the tens place. Its value is four tens, or 40.

5 is in the ones place. Its value is five ones, or 5.

Add the values together to find the value of the whole number.
7000 + 300 + 40 + 5 = 7345

1) Write the number shown by each place value house and partition it. One has been done for you.

Thousands	Ones		
	O	H	T	O
1	4	5	2

$1452 = 1000 + 400 + 50 + 2$

a)

Thousands	Ones		
	O	H	T	O
6	3	4	8

b)

Thousands	Ones		
	O	H	T	O
9	5	6	4

c)

Thousands	Ones		
	O	H	T	O
3	7	1	5

d)

Thousands	Ones		
	O	H	T	O
2	2	8	5

e)

Thousands	Ones		
	O	H	T	O
7	4	9	1

2) Amman makes the number 7045 using these place value arrow cards.

Make these four-digit numbers using place value arrow cards. Draw the place value arrow cards you use each time.
a) 3042   b) 7290   c) 5809   d) 2760
e) 3800   f) 4050   g) 9003   h) 2011

3) Mark Isla's work. For each incorrect answer, explain why she is wrong and give the correct answer.
a) $4000 + 200 + 70 + 9 = 4279$ b) $5000 + 700 + 40 + 5 = 5745$
c) $6000 + 100 + 90 + 4 = 794$ d) $9000 + 40 + 9 = 949$
e) $7000 + 700 + 7 = 7770$ f) $1000 + 1 = 1001$

Let's learn

We can partition the same number in different ways. For example, the number 125 can be:

one hundred	+	two tens	+	five ones	
= 100	+	20	+	5	= 125

We know that 10 tens = 1 hundred (10 × 10 = 100) so the number 125 can also be:

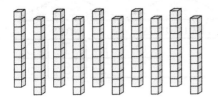 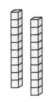

10 tens	+	two tens	+	five ones
= 12 tens	+	five ones		
= 120	+	5		= 125

We know that 10 ones = one ten (10 × 1 = 10) so the number 125 can also be:

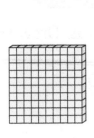 

one hundred	+	20 ones	+	five ones
= one hundred	+	25 ones		
= 100	+	25		= 125

The number 125 can also be 125 ones!

4) Match each number up with two different ways to make it. For example, 263 matches up with two hundreds, five tens and 13 ones.

It also matches 200 + 50 + 13

a) 263

17 tens and five ones

400 + 90 + 10

b) 175

two hundreds five tens and 13 ones

200 + 97

c) 500

two hundreds 20 tens and eight ones

200 + 50 + 13

d) 408

two hundreds and 97 ones

100 + 10 + 10

e) 120

four hundreds nine tens and 10 ones

200 + 200 + 8

297

one hundred one ten and 10 ones

170 + 5

CHALLENGE!

Class 5B is challenged to predict what will happen when they type 9999 + 1 = into a calculator. Try it! We say ten thousand. We write 10 000. A small space separates the thousands from the hundreds, tens and ones. Try writing these numbers in numerals:

Don't forget to put the small space in the correct place.

a) Twenty thousand, four hundred and thirty-nine

b) Sixty-two thousand, one hundred and twenty

2 Number – order and place value

2.4 Comparing and ordering numbers in the range 0–10 000

We are learning to compare and order four-digit numbers.

Before we start

Isla's teacher has asked her to order these numbers from smallest to largest but Isla doesn't understand what this means.

567, 675, 765, 755, 766, 657, 576

Explain what she should do and why. Write the numbers from smallest to largest.

Place value is used to compare and order numbers.

Let's learn

To compare two four-digit numbers, we look at the thousands first. For example, in the numbers **3539** and **9035**, 3539 has three thousands. 9035 has nine thousands. 9035 is the larger number.

The numbers **9309** and **9093** both have nine thousands. We need to look at the hundreds digit to compare them. 9309 has three hundreds. 9039 has no hundreds. 9309 is the larger number.

If the numbers we are comparing have the same thousands **and** hundreds digits then we need to look at the tens digit. If the tens digits are also the same, we need to look at the ones digit. For example, **3009** has nine ones. **3003** has three ones. 3009 is the larger number.

We can compare numbers and arrange them in order. From largest to smallest the numbers in bold are:

9309, 9093, 9035, 3539, 3009, 3003

1) Write each of these sets of numbers in order from smallest to largest.
   a) 2456   2718   2345
   b) 4132   4287   4310
   c) 6014   6521   6375

2) Write each of these sets of numbers in order from smallest to largest and then from largest to smallest:
   a) 3672   3762   3627   3726
   b) 8918   8928   8913   8925
   c) 7384   7374   7372   7381

3) Use these digit cards:
   a) How many different four-digit numbers between 7000 and 8000 can you make?
   b) Write them in order from smallest to largest.
   c) How many different four-digit numbers between 8000 and 9000 can you make?
   d) Write them in order from largest to smallest.

$$\boxed{4}\ \boxed{8}\ \boxed{2}\ \boxed{7}$$

CHALLENGE!

I think 7899 is more than 8700.

I think 7899 is less than 8700.

Who is correct? Explain by writing down your thinking.

2.5 Reading and writing decimal fractions

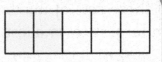
We are learning to read and write decimal fractions with tenths.

Before we start

Amman and Nuria agree that the diagram shows four-tenths but are arguing over how to represent it as a fraction. Amman writes $\frac{4}{10}$. Nuria writes $\frac{10}{4}$. Who is correct? Explain.

A decimal fraction is similar to a fraction. It represents part of a whole.

Let's learn

We can write one-tenth as a fraction and as a decimal fraction.

$\frac{1}{10}$
0·1

A decimal point separates the whole numbers from the parts. There are no whole rectangles and seven-tenths of a rectangle. We write **0·7** and we say **zero point seven**.

$\frac{7}{10}$
0·7

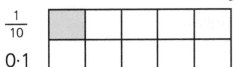

Some decimal fractions have a whole number and a fraction part. The diagram below shows one whole and three-tenths. We write **1·3** and we say **one point three**.

Let's practise

1) Write the decimal fraction shown by each diagram in three different ways. One has been done for you.

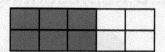

The diagram shows six-tenths. We write **0·6** and we say **zero point six**.

a)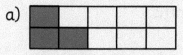

b)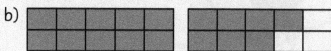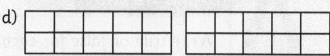

c)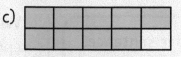

d)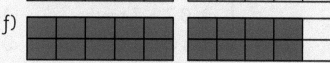

e)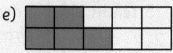

f)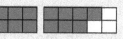

2) Write the decimal fraction shown by each diagram.

Explain why each diagram represents the decimal fraction you have written.

a)

b)

c)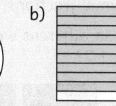

d)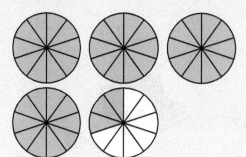

e)

⭐ **CHALLENGE!** ....

You will need a sheet of blank ten-frames. Take turns to spin a 1–9 spinner. Colour that number of tenths. After six rounds, write a decimal fraction to represent the total shaded. The person with the largest decimal fraction is the winner.

2.6 Representing and describing decimal fractions

We are learning to build and describe decimal fractions with tenths.

Before we start

Write true or false for each number sentence.

a) 0·7 = 7 tenths

b) 1·5 = 15 tenths

c) 4·0 = 4

d) 0·2 = 2

e) 2·1 = twenty-one

We can represent decimal fractions in different ways.

Let's learn

Most numbers can be represented as decimal fractions.

We can model the same decimal fraction in different ways. Why does each of the models below represent 1·2?

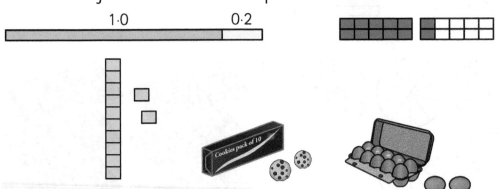

We can describe decimal fractions in numerals and words. For example, 1·2 and **one point two.**

Let's practise

1) Model each decimal fraction in two different ways using fraction tiles, ten-frames or any other resource. Explain your models to a friend.

a) 1·8   b) 2·5   c) 4·3   d) 5·9

2) Play 'Frame the Fraction' with a partner. You will each need a 'window' measuring 5 squares × 2 squares.

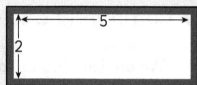

- Take turns to try and 'frame' as many green squares as possible with your fraction window.
- You score points by correctly reading and writing the decimal fraction you have framed. For example, if you frame the ten squares in the top right hand corner of the grid you will have framed five tenths. Writing 0·5 and saying **zero point five** will earn you five points.
- Green squares may be used more than once but the exact same frame cannot be repeated.
- The winner is the player with the most points after an agreed number of rounds or an agreed length of time.

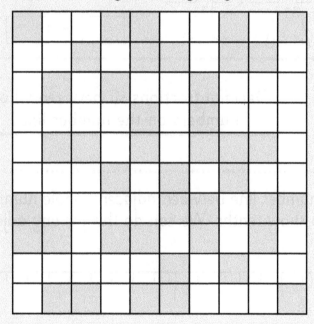

⭐ **CHALLENGE!**

Amman thinks that $\frac{1}{2}$ is the same as 0·2. Isla thinks $\frac{1}{2}$ is the same as 0·5. Who is correct? Explain.

2 Number – order and place value

2.7 Comparing and ordering decimal fractions

We are learning to count on and back in tenths, and place decimal fractions with tenths on a number line.

Before we start

Use the numbers on the stars to make each number statement true. How many different ways can you find?

☐ < ☐

☐ > ☐

☐ ≠ ☐

731   317   171

Decimal fractions sit between whole numbers on the number line.

Let's learn

We can split the number line between adjacent whole numbers into ten equal parts to show tenths. We can do this for any adjacent whole numbers.

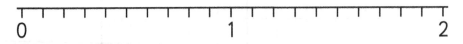

0   1   2

We can compare decimal fractions using the symbols 'is equal to' (=), 'is less than' (<) and 'is greater than' (>). If two decimal fractions share the same whole number we only need to compare the tenths parts to say which decimal fraction is larger or smaller.

0·8 < 0·9 because eight tenths is less than nine tenths.

1·6 > 1·3 because six tenths is more than three tenths.

Let's practise

1) Write the decimal fractions that are missing from these number lines.

a)
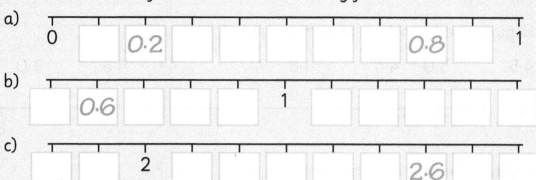

0 ☐ 0.2 ☐ ☐ ☐ ☐ 0.8 ☐ 1

b)
☐ 0.6 ☐ ☐ ☐ 1 ☐ ☐ ☐ ☐ ☐

c)
☐ ☐ 2 ☐ ☐ ☐ ☐ ☐ 2.6 ☐

2) Choose the correct symbol to make each statement true. > < =

a) 0·8 ☐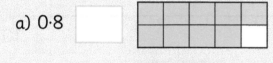

b) 0·3 ☐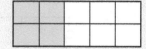

c) 0·7 ☐

d) 0·9 ☐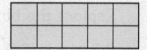

e) 0·4 ☐

f) 0·6 ☐

3) Copy and complete each number line. Use the completed number lines to help you decide if you should use <, > or = to make each statement true.

a)

0 ☐ ☐ ☐ ☐ ☐ ☐ ☐ ☐ 1

Replace the box with the correct symbol, <, > or =.

0·2 ☐ 0·5   0·4 ☐ 0·3   0·7 ☐ 0·5   0·9 ☐ 1·0

b)
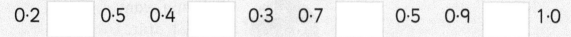

2 ☐ ☐ ☐ ☐ ☐ ☐ ☐ ☐ 3

Replace the box with the correct symbol, <, > or =.

2·5 ☐ 2·9   2·7 ☐ 2·3   2·6 ☐ 2·6   2·8 ☐ 2·4

c)

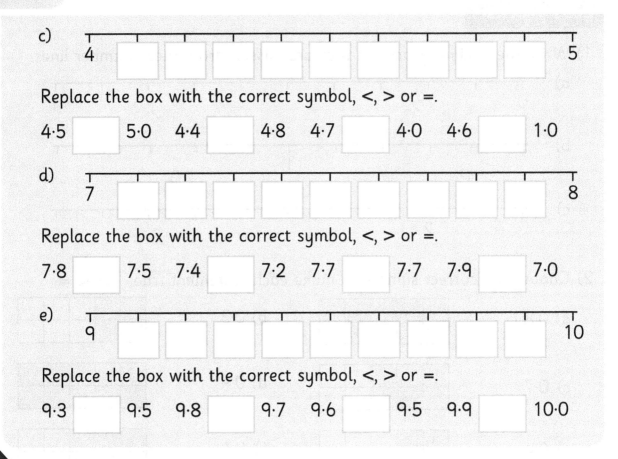

4 |□|□|□|□|□|□|□|□|□| 5

Replace the box with the correct symbol, <, > or =.

4·5 [ ] 5·0   4·4 [ ] 4·8   4·7 [ ] 4·0   4·6 [ ] 1·0

d)

7 |□|□|□|□|□|□|□|□|□| 8

Replace the box with the correct symbol, <, > or =.

7·8 [ ] 7·5   7·4 [ ] 7·2   7·7 [ ] 7·7   7·9 [ ] 7·0

e)

9 |□|□|□|□|□|□|□|□|□| 10

Replace the box with the correct symbol, <, > or =.

9·3 [ ] 9·5   9·8 [ ] 9·7   9·6 [ ] 9·5   9·9 [ ] 10·0

⭐ **CHALLENGE!** ⬤⬤⬤⬤⬤⬤⬤⬤⬤⬤⬤⬤⬤⬤⬤⬤⬤⬤⬤

Nuria buys a bar of chocolate to share with her friends. Use the words in the speech bubbles to work out the answers:

a) Who gets the least chocolate? Explain.

b) Who gets the most chocolate? Explain.

I only want 0·1 of it.

Amman

I will take the rest and give Finlay 0·4.

I'm going to eat three tenths of the chocolate bar.

Nuria

Isla

Number – order and place value

2.8 Recognising the context for negative numbers

We are learning to recognise where negative numbers are used in the environment.

Before we start

Write the number each arrow is pointing to.

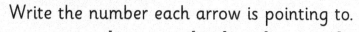

2 000   3 000   4 000

There are numbers below zero called negative numbers.

Let's learn

Negative numbers are whole numbers below zero. Negative numbers are usually written with a minus sign in front. For example, −3 means **minus three** or **negative three**.

Whole numbers above zero are called positive numbers. Positive numbers are sometimes written with a plus sign in front. For example, +3 means **plus three** or **positive three**.

Zero is neither positive nor negative.

This weather map shows the temperature in Dundee as **zero degrees Celsius (0 °C).**

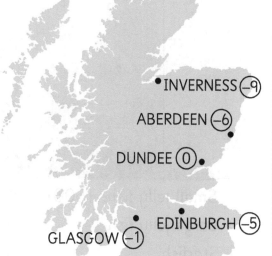

•INVERNESS ⊖9

ABERDEEN ⊖6 •

DUNDEE ⓪ •

EDINBURGH ⊖5 •

GLASGOW ⊖1 •

Let's practise

1) Use the map on the previous page. Write the temperature in these cities in three ways. One has been done for you.

   a) Inverness: Nine degrees below zero. Minus nine degrees Celsius. $-9\,°C$

   b) Aberdeen

   c) Edinburgh

   d) Glasgow

2) Record the depth above or below sea level for each creature in the picture. One has been done for you.

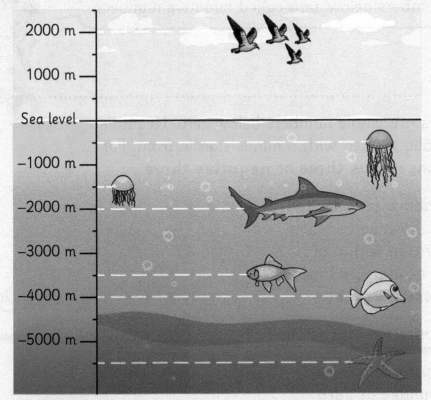

   a) Small jelly fish: 1500 m below sea level

   b) Large jelly fish   c) Orange fish   d) Shark

   e) Starfish   f) Yellow fish   g) Birds

3) In golf, a minus score means the player is under par (has taken less shots than expected). A plus score means the player is over par (has taken more shots than expected). This table shows the leaderboard after two rounds of the local golf championship.

Leaderboard after two rounds	
McIntosh	–5
Brown	–3
Smith	–2
Gray	–1
Morrison	+1
Lyle	+2
Murray	+4
Gordon	+6

Write each player's score using the words 'under par' or over par'. One has been done for you.

McIntosh: 5 shots under par.

CHALLENGE!

Investigate to find more examples of where negative numbers are used. How are they used?
What do they mean?

3.1 Mental addition and subtraction

> We are learning to add and subtract whole numbers mentally.

Calculate mentally. Explain how you worked each answer out.
a) 39 + 39   b) 8 + 537   c) 120 – 60   d) 258 – 9

> If you **compensate** for something, you do something to make up for it. Compensation helps us to add and subtract numbers mentally.

Let's learn

Round and adjust:

To find 273 + 29, we might round 29 up to 30. 273 + 30 = 303 but we have added one too many. We need to **compensate** for this by subtracting one. 303 – 1 = 302 so the answer is 302. How might we use round and adjust to find 477 – 59?

Give and take:

We can take some from one number and **compensate** for this by giving it to the other number. Let's calculate 47 + 28. Subtracting 2 from 47 and adding it to 28 will give us 45 + 30 which is an easier calculation. Can you find 47 + 28 by giving and taking 3?

Let's practise

1) Use 'round and adjust' to work out the answers to these calculations.

> The question is 87 – 49. Round 49 up to 50. 87 minus 50 is 37. Compensate for taking away one too many by adding one back on. 37 + 1 = 38 So 87 – 49 = 38

a) 87 – 49   b) 63 + 19   c) 155 – 39
d) 218 + 29   e) 696 – 59   f) 253 + 28
g) 173 – 18   h) 257 + 38   i) 494 – 48

2) Use 'give and take' to work out the
answers to these calculations.

a) 57 + 38   b) 24 + 46
c) 17 + 28   d) 62 + 68
e) 77 + 48   f) 197 + 98
g) 814 + 36   h) 413 + 61
i) 51 + 333   j) 22 + 448

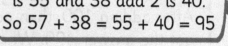

The question is 57 + 38.
I can take 2 from 57 and
add it to 38. 57 minus 2
is 55 and 38 add 2 is 40.
So 57 + 38 = 55 + 40 = 95

3) | 125 54 66 78 64 | | 19 21 29 131 38 |

a) Using one number from the yellow box and one number from
the green box, write five **different addition questions** and five
different subtraction questions.

b) Calculate each answer using a strategy of your choice.

c) Explain to a partner how you worked each answer out.

CHALLENGE!

Isla and Finlay are describing how they used compensation.

a) What was Isla's question? Write her question and answer it.

52 is near to 50, so I rounded it down to 50. 31 add 50
is 81, so I just needed to add another 2 on.

b) What was Finlay's question? Write his question and answer it.

First I rounded the second number up to 40. I know that
56 take away 40 is 16, but this was taking away one too
many so I added 1 to the answer.

3 Number – addition and subtraction

3.2 Adding and subtracting 1, 10, 100 and 1000

We are learning to use place value to add and subtract 1, 10, 100 or 1000 to and from four-digit numbers.

Before we start

Amman wrote a three-digit number on his mini whiteboard. After adding 100, then adding 10, then subtracting 1 he got the answer 499. What number is written on Amman's mini whiteboard?

The patterns in our place value system can help us with addition and subtraction.

Let's learn

We can count by ones, tens, hundreds or thousands from any four-digit number. Try reading the numbers on these number lines, forwards and backwards. By how much are the numbers increasing or decreasing each time?

5070   5080   5090   5100   5110   5120   5130   5140   5150   5160

3600   3700   3800   3900   4000   4100   4200   4300   4400   4500

Let's practise

1) Write the next five numbers in each sequence.

   a) 1124, 1125, 1126 ...   b) 4693, 4694, 4695 ...
   c) 8996, 8997, 8998 ...   d) 2875, 2874, 2873 ...
   e) 5203, 5202, 5201 ...   f) 7006, 7005, 7004 ...

2) a) Add 1 onto each number:

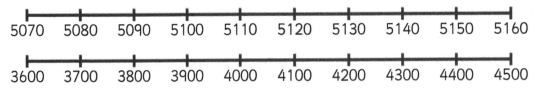

   4949,   6780,   1599,   3999

b) Subtract 1 from each number:

6020,   8300,   2093,   5000,

3) a) Start on the bottom rung of each ladder and keep adding 100 until you reach the top.

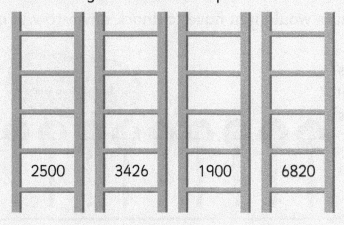

2500   3426   1900   6820

b) Start on the top rung of each ladder and keep subtracting 1000 until you reach the bottom rung.

9999   7081   4200   4001

★ CHALLENGE! ....

1) Nuria is thinking of the number 3790. Write the number that is:
   a) one more   b) 10 more   c) 100 more   d) 1000 more
   e) one less   f) 10 less   g) 100 less   h) 1000 less

3.3 Adding and subtracting multiples of 100

We are learning to add and subtract multiples of 100.

Before we start

Which coconuts would you have to knock down to win a prize in:

a) two shots?
b) three shots?
c) four shots?

Five shots for £2
Score 300 to win!

180 120 50 70 140 60 30 100 150

We can use number bonds and place value to add and subtract multiples of 100.

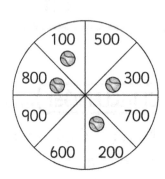

Let's learn

Finlay knows that 8 + 3 + 1 + 2 = 14

The numbers 80, 30, 10 and 20 are ten times bigger than 8, 3, 1 and 2. So 80 + 30 + 10 + 20 = 140

The numbers 800, 300, 100 and 200 are one hundred times bigger than 8, 3, 1 and 2. So 800 + 300 + 100 + 200 = 1400

We can use a similar strategy to subtract multiples of 100. For example,

$$14 - 3 - 1 - 2 = 8$$

So $140 - 30 - 10 - 20 = 80$

and $1400 - 300 - 100 - 200 = 800$

100 500
800 300
900 700
600 200

Let's practise

1) Nuria   Isla   Amman

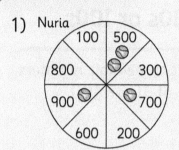  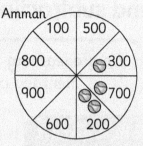

a) Who scored the same number of points as Finlay (1400)?

b) Who scored the most points? What was their score?

c) Write the scores in order from highest to lowest.

d) Find three different ways to score exactly 2000 using four balls.

2) Copy and complete the patterns.

a)

9	+	3	+	1	+	8	=	☐
☐	+	30	+	10	+	80	=	☐
900	+	☐	+	100	+	800	=	☐

b)

34	−	10	−	8	−	5	=	☐
340	−	100	−	80	−	☐	=	☐
3400	−	☐	−	☐	−	500	=	☐

c)

56	−	6	−	5	−	9	=	☐
560	−	☐	−	50	−	90	=	☐
5600	−	600	−	☐	−	☐	=	☐

CHALLENGE!

| 400 | 300 | 800 | 600 | 700 | 900 | 200 | 500 |

a) Hook five ducks to make the **highest** score possible.

b) Hook five ducks to make the **second lowest** score possible.

c) Find the difference between your answers to questions a) and b).

3.4 Adding and subtracting by making 10s or 100s

We are learning to add and subtract three-digit numbers mentally by making multiples of 10 or 100.

Before we start

Ms McFarlane has challenged Finlay to see how many addition and subtraction facts he can write for the number 100 but he has made some mistakes. Correct his work. For each incorrect answer, write what Finlay should have written.

a) 64 + 36 = 100   b) 29 + 81 = 100
c) 55 + 55 = 100   d) 100 – 27 = 83

Looking for multiples of 10 or 100 can make mental addition and subtraction easier.

Let's learn

We can rearrange the order of an addition to make the calculation easier. One way to do this is to look for pairs of numbers that go together to make a multiple of 10 or 100 first, then partition the remaining number. For example:

| 57 + 126 + 33 |

57 + 33 = 90; 90 + 100 = 190; 190 + 20 = 210; 210 + 6 = 216

| 79 + 350 + 150 | 350 + 150 = 500;   500 + 79 = 579

When subtracting, it helps to think about pairs of numbers too. For example:

| 900 – 648 | 48 'fits with' 52 to make 100.

648 + **52** = 700   700 + **200** = 900 so the answer is 252.

Let's practise

Amman uses jottings to help him keep track of his thinking.

The question is 262 + 29 + 31

29 add 1 makes 30, plus another 30 makes 60. 60 plus 60 is 120, so that's 200 plus 120 plus 2...makes 322.

Jottings

262 + 60
200 + 120 + 2
322

1) Calculate mentally. Look for multiples of 10 or 100. You may find it helpful to jot these numbers down.

   a) 22 + 717 + 48
   b) 35 + 65 + 400
   c) 210 + 26 + 490
   d) 128 + 463 + 37
   e) 51 + 111 + 669
   f) 145 + 215 + 355

2) Finlay imagines an **empty number line** to help him solve 800 − 424. He counts on to the next multiple of 10, then to the next multiple of 100. Finally, he adds the remaining 100s.

I can picture a number line in my head but I will jot down 6, 70 and 300 to remind me to add them together.

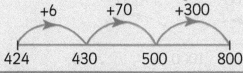

Calculate mentally. You may find it helpful to imagine counting on or back on an empty number line and jot down the jumps you make.

   a) 700 − 324
   b) 400 − 138
   c) 600 − 219

3) Now find the missing number in each of these calculations in only two jumps, by imagining or drawing an empty number line.

   a) 800 − ☐ = 674   b) ☐ + 189 = 500   c) 1000 − 281 = ☐

★ **CHALLENGE!** ·······························

Fill in the missing digits. Write the completed number sentences.

4✳3 + 6✳ = 500   600 − 24✳ = 3✳5

3.5 Adding and subtracting multiples of 1000

We are learning to add and subtract numbers to and from multiples of 1000.

Before we start

Replace each ? with a multiple of 10 so that each row totals 1000.

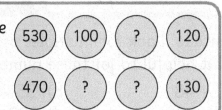

530 100 ? 120

470 ? ? 130

When adding or subtracting, an empty number line can help us keep track of our thinking.

Let's learn

Nuria and Isla are working out the answer to 234 + ☐ = 2000.

Isla counts on from 234 up to 2000

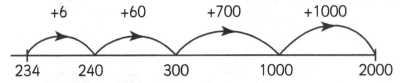

+6   +60   +700   +1000

234   240   300   1000   2000

Nuria counts back from 2000 to 234.

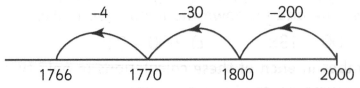

−4   −30   −200

1766   1770   1800   2000

Amman is working out the answer to 5000 − 1472 = ☐

He counts back from 5000 then checks his answer by counting on.

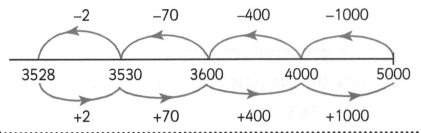

−2   −70   −400   −1000

3528   3530   3600   4000   5000

+2   +70   +400   +1000

Let's practise

1) Work out the missing numbers on each number line, then write one addition and one subtraction to match it. One has been done for you.

a)

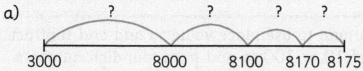

```
?        ?      ?    ?
3000           8000      8100  8170 8175
```

The missing numbers are 5000, 100, 70 and 5.
3000 + **5175** = 8175 8175 − **5175** = 3000

b)

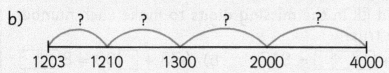

```
?       ?       ?        ?
1203   1210    1300     2000      4000
```

c)

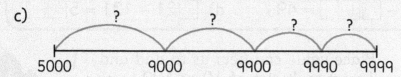

```
?        ?      ?   ?
5000        9000     9900   9990 9999
```

2) Choose between counting on and counting back to solve these problems.

a) ☐ + 1468 = 8000

b) 2222 + ☐ = 6000

c) 9000 = 7430 + ☐

d) ☐ − 1305 = 4000

e) 2000 − ☐ = 555

f) 3000 − 1561 = ☐

⭐ **CHALLENGE!**

Complete the puzzles by writing a four-digit number in each empty triangle. Triangles round the outside must total the number in the centre.

a)

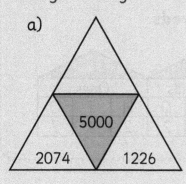

5000

2074   1226

b)

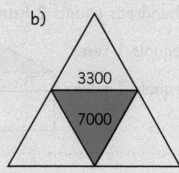

3300

7000

c)

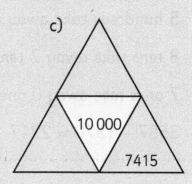

10 000

7415

3.6 Adding and subtracting multiples of 10 and 100

We are learning to use place value to add and subtract multiples of 10 or 100 to and from four-digit numbers.

Before we start

Copy and fill in the missing digits to make each number sentence true:

a) 417 + ☐☐☐ = 817

b) 673 + ☐00 = 8☐☐

c) 349 – ☐☐☐ = 49

d) ☐21 – 121 = 5☐☐

Place value can help us to add and subtract multiples of 10 or 100 to and from a four-digit number.

Let's learn

Nuria uses **place value partitioning** to work out 2335 + 190.

She splits 2335 into **2000 + 300 + 30 + 5**, and 190 into **100 + 90**.

2000 + 300 + 100 + 30 + 90 + 5 + 0 =
2000 + 400 + 120 + 5 = 2525.

Finlay uses **number facts and place value** to work out 3587 – 1370.

3 thousands take away 1 thousand equals **2 thousands**.

5 hundreds take away 3 hundreds equals **2 hundreds**.

8 tens take away 7 tens equals **1 ten**.

7 ones take away 0 ones equals **7 ones**.

3587 – 1370 = 2217

Thousands	Ones		
O	H	T	O
2	2	1	7

Let's practise

1) Use Nuria's strategy to work out:

   a) 7324 + 180   b) 6295 + 330   c) 5251 + 1500

   d) 2877 + 3200   e) 6794 − 760   f) 3875 − 520

   g) 4699 − 1470   h) 8858 − 3250   i) 9786 − 333

2) Use Finlay's strategy to work out:

   a) 9563 + 120   b) 7658 + 240   c) 2385 + 4400

   d) 5274 + 3710   e) 2657 − 330   f) 4795 − 510

   g) 8230 − 6120   h) 6779 − 3250   i) 9127 − 8005

3) Replace ♥ with a multiple of 10 or 100 to make each number sentence true.

   a) 2335 + ♥ = 2935   b) 5922 + ♥ = 5962

   c) 1371 + ♥ = 1691   d) 3587 − ♥ = 3517

   e) 8529 − ♥ = 8329   f) 6745 − ♥ = 6605

CHALLENGE!

Isla asks her friends the following question.

What is 3243 minus 1800?

Finlay thinks the answer to Isla's question is 2643.

Explain why Finlay is incorrect.

Work out the correct answer to Isla's question.

You may find it helpful to draw an empty number line.

3.7 Solving word problems

We are learning to choose the most efficient strategy to solve addition and subtraction problems.

Before we start

Identify the missing numbers in these bar models. They are not drawn to scale.

500		
	53	240

800		
95		95

Bar models help us to solve word problems by focusing on the parts that make up the whole.

Let's learn

Nuria and Isla are working together on this problem:

> Jan is saving up for a new bike costing £220. So far, she has saved £93. She gets £50 from her gran for her birthday. How much money does Jan still need?

The girls decide to draw a bar model to help them to think about the problem.

220		
93	50	?

$143 + 7 = 150$

$150 + 50 = 200$

$200 + 20 = 220$

Jan still needs £77.

Isla adds 50 to 93 and jots down 143. Then she counts on from 143 up to 220.

Nuria solves the problem by drawing an empty number line.

Talk with a partner about their strategies. How would you solve the problem?

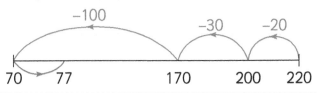

Let's practise

Draw a bar model to represent each problem then solve the problem using a strategy of your choice.

1) The Big Wheel had 5260 visitors between Monday and Friday and 1740 more visitors at the weekend. How many people were on the Big Wheel over the whole week?

2) On Friday, 1200 people went on the Dodgem Cars, 1526 went on the Octopus and 1300 went on the Waltzers.
   a) How many people is this altogether?
   b) Find the difference between the most and least popular ride.

3) 2500 people went on the Ghost Train on Saturday. 740 of them were children. How many adults were on the Ghost Train?

4) The Roller Coaster can take 212 people. 98 people are on it. How many more could get on?

5) a) 2356 visitors visited the fairground on Saturday and 2897 visited on Sunday. How many fewer visitors were there on Saturday?
   b) 9000 people visited the fairground over the whole week. How many visitors were there from Monday to Friday?

CHALLENGE!

Write a word problem to match this bar model. Ask a friend to solve your word problem.

876		
137	239	?

3.8 Representing word problems in different ways

We are learning to represent the same word problem in different ways.

Before we start

Represent this word problem as a bar model and on an empty number line. Write a number sentence to match how you solved it.

Isla's mum has 466 letters in her sack. She has already delivered 371 letters. How many letters did she have at the start of her round?

Representing the same problem in different ways helps us to understand it better.

Let's learn

Amman and Finlay are working on this problem:

The school library has 1307 books in it. The librarian only has 854 bookmarks. How many books will not have a bookmark?

The boys draw a Think Board about the problem.

Word problem	Bar model
The school library has 1307 books in it. The librarian only has 854 bookmarks. How many books won't have a bookmark?	1307 ? 854
Empty number line −50 −100 −300 −7 850 +4 900 1000 1300 1307 854	**Number sentence** 1307 − $\boxed{453}$ = 854

1) Create Think Boards for these word problems and solve them. Represent each problem as a bar model, as an empty number line and as a number sentence.

   a) In a TV dancing contest, 5356 people voted for Couple Number One, 3206 people voted for Couple Number Two and 6864 voted for Couple Number Three. Find the difference between the highest and lowest scores.

   b) Isla's mum delivered 2768 leaflets between Monday and Thursday. How many leaflets does she still need to deliver if she started with 3200?

   c) Finlay and his mum are at a football match. There are 868 spectators in the north stand. The stand can hold 2000 people. How many empty seats are there?

2) It is approximately 339 km by road from Ullapool to Edinburgh, and approximately 1000 km by road from Ullapool to London. How much further is it to London than Edinburgh?

   Nuria says: 'I think this is a subtraction question.'

   Isla says: 'I think this is an addition question.'

   a) Who is correct, Nuria or Isla? Explain why you think this.

   b) Create a Think Board for this problem and solve it.

⭐ CHALLENGE! ....

Write your own word problems with at least one four-digit number in them. Challenge your friends to create Think Boards for your problems and solve them.

3.9 Using non-standard place value partitioning

> We are learning to partition numbers in different ways.

Before we start

Candy canes are sold singly or in packs of 10.

Find **all** the different ways that Ms Higgins can buy **68** candy canes for the P5 party. One has been done for you.

68 can be 5 packs of 10 and 18 single candy canes.
50 + 18 = 68

> We can partition three-digit numbers in different ways by **exchanging** hundreds for tens.

Let's learn

Amman and Isla are investigating different ways to make the number 461 using place value counters. They know that 400 + 60 + 1 = 461

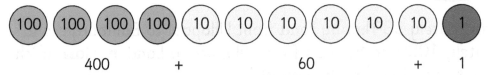

$$400 \quad + \quad 60 \quad + \quad 1$$

Amman says: 'We know that 10 tens equal 100. If we exchange one of the hundreds for 10 tens, we will have 3 hundreds, 16 tens and 1 one. That's 300 + 160 + 1.'

Isla says: 'We can exchange two hundreds for 20 tens, and then we will have 200 + 260 + 1.'

Exchanging three hundreds counters for 30 tens counters would give Amman and Isla 1 hundred, 36 tens and 1 one = 100 + 360 + 1.

Exchanging all four hundreds counters for 40 tens counters would give them 46 tens and 1 one = 460 + 1.

Let's practise

1) Candy canes can be bought in boxes of 100 as well as packs of 10 and as singles.

Ms Higgins decides to buy enough candy canes for the P6 party, too. Show four different ways that she could buy **135** candy canes. Include a box of 100 each time.

Box of 100

2) Write the total represented by each set of place value counters in two different ways. One has been done for you.

a) (100)(100)(10)(10)(10)(10)(10)(10)(10)(10)(10)(1)(1)
(10)(10)(10)(10)(10)(10)(10)(10)(10)

382 = 2 hundreds, 18 tens and 2 ones = 200 + 180 + 2

b) (100)(100)(100)(10)(10)(10)(10)(10)(10)(1)(1)(1)(1)
(10)(10)(10)(10)(10)(10)(1)(1)(1)

c) (100)(100)(100)(10)(10)(10)(10)(10)(10)(1)(1)(1)(1)
(100)(100)(100)(10)(10)(10)(10)(10)

CHALLENGE!

a) Find three ways to make 756 by exchanging tens counters for ones counters, for example:

(100)(100)(100)(100)(10)(10)(1)(1)(1)(1)(1)(1)(1)(1)
(100)(100)(100)   (10)(10)(1)(1)(1)(1)(1)(1)(1)(1)

756 = 700 + 40 + 16

b) Can you make 756 by exchanging one hundreds counter **and** one tens counter?

c) How many other ways can you find to make 756 using place value counters?

3.10 Adding using a semi-formal written method

> We are learning to use column addition.

Before we start

This set of place value counters shows 700 + 50 + 6 = 756.

(100) (100) (100) (100) (10) (10) (10) (10) (10) (1) (1) (1) (1)

(100) (100) (100)   (1) (1)

$$700 \quad + \quad 50 \quad + \quad 6 \quad = 756$$

Find six other ways to make the number 756 by exchanging hundreds counters for tens counters.

> Some numbers are easier to add together if we set them down in a column.

Let's learn

Writing numbers down in columns can help us add those that are tricky to work with mentally. This is because columns make it easier to look at the ones, the tens and the hundreds separately. For example, 576 + 268 can be written as:

```
                        576
                      + 268
Add the hundreds        700
Add the tens            130
Add the ones             14
Find the total          844
```

> 500 add 200 is 700
> 70 add 60 is 130
> 6 add 8 is 14
> 700 + 130 + 14 = 844

Let's practise

1) Use Nuria's method to calculate.

    a) 645 + 138   b) 245 + 526   c) 417 + 446

    d) 324 + 238   e) 772 + 183   f) 182 + 284

    g) 375 + 354   h) 274 + 546   i) 486 + 354

    j) 175 + 579   k) 266 + 766   l) 848 + 484

2) Now find the answers to these additions using the column method.

    a) 373 + 113 + 414   b) 552 + 327 + 784   c) 303 + 555 + 62

3) Use the numbers on the cards to make each number sentence true. You may use each card only once.

    a) 5368 + ☐ − ☐ = 5838

    b) 6189 − ☐ + ☐ = 6449

    c) 2035 − ☐ − ☐ + ☐ = 1715

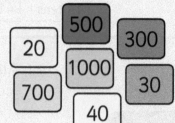

500   300   20   1000   700   30   40

⭐ CHALLENGE!

A **palindrome** is a number that reads the same forwards and backwards, for example 474. Find the missing digit so that each total is a palindrome.

a) 200 + ◆50 + 3   b) 400 + ◆60 + 9

c) ◆00 + 320 + 8   d) ◆00 + 210 + 5

3.11 Adding three-digit numbers using standard algorithms

> We are learning to add three-digit numbers using a standard written method.

Before we start

Use two or more cards to make the number 149 in three different ways. Cards may be used more than once. Not all cards may be needed.

1 hundred	19 tens	94 tens	49 ones

94 ones	4 tens	4 ones

14 tens	41 tens	9 ones

> Algorithms can help us with addition calculations that are too difficult to work out mentally.

Let's learn

An algorithm is a set of instructions for carrying out a calculation. The steps must be carried out in the correct order.

Add the ones. 5 ones + 6 ones = 11 ones = **1 ten** and 1 one. Write 1 in the ones column and **carry** the **1 ten** over into the tens column.

Add the tens. **1 ten** + 3 tens + 8 tens = 12 tens = **1 hundred** and 2 tens. Write 2 in the tens column and **carry** the **1 hundred** over into the hundreds column.

Add the hundreds. **1 hundred** + 2 hundreds + 3 hundreds = 6 hundreds

```
  1 1
  2 3 5
+ 3 8 6
  6 2 1
```

Let's practise

1) Write each calculation as a column addition. Use Amman's method to find the answers.

   a) 196 + 733   b) 408 + 362   c) 556 + 271
   d) 868 + 490   e) 379 + 383   f) 426 + 479
   g) 627 + 294   h) 569 + 783   i) 648 + 175
   j) 773 + 855   k) 949 + 606   l) 246 + 688

2) Calculate the answers to these questions using the standard written algorithm for addition.

   a) 126 + 545 + 379   b) 788 + 243 + 913
   c) 434 + 577 + 679   d) 616 + 414 + 55

3) a) Use each digit only once each time to write a column addition where you need to:

   3 7 2 8 5 0

   i) carry a ten over into the tens column
   ii) carry a hundred over into the hundreds column
   iii) carry over both a ten and a hundred.
   b) Ask a partner to find the answers.
   c) Ask a partner to make up some questions for you to solve.

⭐ **CHALLENGE!**

Copy and fill in the missing digits.

```
   2 4 ✳           ✳ 0 8
 + ✳ 8 2         + 6 ✳ 9
 ─────────       ─────────
   6 ✳ 9           1 1 3 ✳
```

3.12 Representing and solving word problems

We are learning to represent and solve word problems.

Before we start

A can of lemonade holds 330 ml. A bottle of lemonade holds 545 ml. Isla wants to fill a 1000 ml container from the can and the bottle. Does she have enough lemonade? Explain your thinking.

A word problem can be solved in more than one way.

Let's learn

Isla and Nuria have been working on this word problem.

> In one week a supermarket sells 239 Gala apples, 357 Jazz apples and some Golden Delicious apples. If 978 apples were sold altogether, how many were Golden Delicious apples?

They drew a bar model to help them think about the problem.

Nuria added 239 and 357 using the addition algorithm. Then drew an empty number line and counted from 596 up to 978.

978		
239	357	?

```
    1
    2 3 9
  + 3 5 7
    5 9 6
```

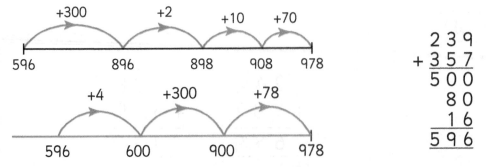

```
    2 3 9
  + 3 5 7
    5 0 0
      8 0
      1 6
    5 9 6
```

Isla was not confident with the addition algorithm so she partitioned 239 and 357 and added the parts. Isla drew a number line for the second part too. The answer is 382.

Let's practise

Represent and solve these word problems using a method of your choice.

1) Finlay and his family are touring Scotland with their caravan. They travel 133 miles on Tuesday and 146 miles on Wednesday. When they reach Braemar on Thursday they have travelled 446 miles since leaving their house. How far did they travel on Thursday?

2) On Friday they decide to visit the Braemar Highland Games before travelling on to Aviemore. 346 people watch the caber-tossing event and 535 people watch the hammer throwing. How many more people watch the hammer throwing?

3) From Aviemore to the Isle of Skye is 133 miles. When they stop off at Urquhart Castle they still have 88 miles to go. How far is it from Aviemore to Urquhart Castle?

4) a) Urquhart Castle receives 687 visitors on Saturday and 928 visitors on Sunday. How many people visit Urquhart Castle over the weekend?

   b) 549 of the people who visit the castle on Sunday are adults. How many children visit the castle on Sunday?

5) The population of Plockton is 379. This is 453 fewer people than live in Finlay's home town. What is the population of Finlay's home town?

6) On Monday the family travels to Edinburgh to visit the zoo. The gift shop sells some keyrings, 109 tea towels and 262 mugs. If 787 gifts were sold altogether, how many keyrings does the gift shop sell?

CHALLENGE!

a) Write two addition number sentences and two subtraction number sentences to fit this bar model.

916	
239	677

b) Choose one of your number sentences and write a word problem to match it.

4.1 Recalling multiplication and division facts for 2, 5 and 10

We are learning to recall multiplication and division facts for 2, 5 and 10.

Before we start

Nuria has 10 bags of marbles. Each bag contains two marbles. Show the problem using materials or by drawing a picture. Then write a number sentence to work out how many she has altogether.

Recalling multiplication and division facts means knowing them straight away without having to think about them. This knowledge is very useful, because it helps us use strategies and solve problems.

Let's learn

Facts that we already know can help us recall multiplication and division facts for 2, 5 and 10.

If we know doubling and halving facts then we already know multiplication and division facts for 2, such as double 6 = 12, so 2 × 6 = 12.

If we know how many tens are in a decade, then we already know multiplication and division facts for 10, such as nine tens are in 90, so 9 × 10 = 90.

We can also use our knowledge of tens to help us with fives:

For example, if we know there are four tens in 40, then we also know there are eight fives in 40.

Use your knowledge to help you recall these facts as quickly as you can.

1) Can you answer these multiplications within one minute? Time yourself:

a) 2×6 b) 6×5 c) 8×10

d) 9×2 e) 4×5 f) 3×10

2) Multiply these numbers by 2, then 5 and finally 10:

a) 6 b) 4 c) 10 d) 9

3) Copy and complete the following number sentences by finding the number underneath the paint splashes.

a) $\text{✻} \times 2 = 14$ b) $5 \times \text{✻} = 30$ c) $8 \times 10 = \text{✻}$

d) $2 \times \text{✻} = 2$ e) $\text{✻} \times 5 = 45$ f) $\text{✻} \times 10 = 100$

g) $2 \times 9 = \text{✻}$ h) $\text{✻} \times 2 = 16$ i) $\text{✻} \times 5 = 20$

j) $\text{✻} \times 2 = 8$ k) $5 \times \text{✻} = 35$ l) $10 \times \text{✻} = 50$

CHALLENGE!

Remove all the face cards from a set of playing cards and shuffle them. Work with a partner to practise multiplication facts for 2. Take turns to each take a card and give the answer for the 2 times table. For example, if you take a 4, you would give the answer to 4×2.

Listen carefully to see if your partner gives the right answer. Which are the tricky ones?

Next, practise facts for 10, and then facts for 5.

For an extra challenge, see if you can give a division fact to match your multiplication fact. For example, if you are practising the 5 times table and take a 4 card, give the correct answer of $4 \times 5 = 20$, and also $20 \div 5 = 4$.

4.2 Recalling multiplication and division facts for 3

We are learning to recall multiplication and division facts for 3.

Before we start

Make or draw an array to show 5 × 3.
How many altogether?

How would you change the array to show 6 × 3?

Recalling multiplication and division facts means knowing them straight away without having to think about them. This knowledge is very useful, because it helps us use strategies and solve problems.

Let's learn

Try to spot which questions take you longer to recall and then you know which facts you need to practise.

Let's practise

1) a) What do you multiply by 3 to get 18?

   b) 12 divided by 3 = ☐

   c) What is 9 times 3?

   d) How many threes are in 15?

   e) 7 × ☐ = 21

   f) 3 × 10 = ☐

   g) What is 6 split into 3?

h) What do you multiply by 3 to get 24?

i) 3 × 3 = ▢

j) How many threes are in 21?

k) 30 divided by 3 = ▢

l) What do you multiply 3 by to get 27?

2) Use your knowledge of multiplication facts to complete these triangles. The first one has been done for you:

a)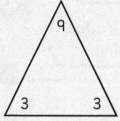

b)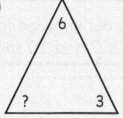

c)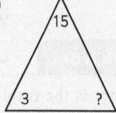

d)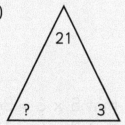

e)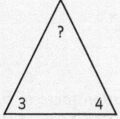

f)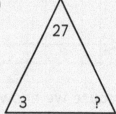

3) Write a division fact for each of the triangles above.

⭐ **CHALLENGE!** ....

Make some number triangles of your own.

Cover up one of the numbers using your thumb. Ask a friend to work out what the answer is by using multiplication or division facts to help.

Ask them to give you some triangle challenges too!

4.3 Solving multiplication problems

We can use facts we know from the 2, 5 and 10 times tables to solve multiplication problems.

Before we start

Finlay bakes muffins and fills a whole tray. The tray has four rows with five muffins in each row.

Make an array or draw a diagram to represent this problem.

How many muffins does he bake altogether?

We can use our knowledge of multiplication facts for 2, 5 and 10 to solve problems.

Let's learn

What is the total of 7 × 6?

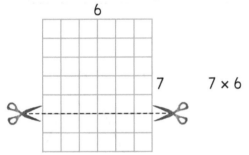

7×6

Because we know multiplication facts of 5, we know that $5 \times 6 = 30$.
Because we know multiplication facts of 2, we know that $2 \times 6 = 12$.

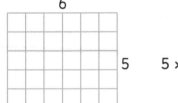

5×6

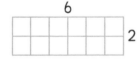

2×6

We use brackets to show how we split up and grouped our calculation:
$7 \times 6 = (5 \times 6) + (2 \times 6)$
$\qquad = 30 + 12$
$\qquad = 42$

The brackets are important because they show which calculations we did first.

1) Use squared paper to make an array for 7 × 7.

   a) What is the answer to 5 × 7? Cut or colour 5 × 7 on your array.

   b) What is the answer to 2 × 7? Cut or colour 2 × 7 on your array.

   c) Now add your answers together to work out 7 × 7.

   d) (5 × 7) + (2 × 7) = __ + __ = __

      Fill in the missing numbers to show how you would record your calculation.

2) A chocolate bar has eight rows of seven squares.

   Use 2, 5 or 10 facts to help you work out how many squares altogether.

   Record your calculation using brackets to show how you split up and grouped the problem.

$$(\underline{} \times \underline{}) + (\underline{} \times \underline{}) =$$

3) How could you use 2, 5 or 10 facts to help you work out these multiplication problems?

   Draw arrays to help you. Use brackets to record your calculations.

   a) 4 × 7   b) 6 × 6   c) 6 × 7

   d) 4 × 6   e) 12 × 7   f) 15 × 6

CHALLENGE!

Use multiplication facts to help you work out this problem.

At the supermarket, there are six rows of 25 cans of beans. Help Isla work out how many there are altogether for the annual stock take.

What facts could you use to help you?

Make an array or draw a diagram to help you work it out.

Use the bracket notation you have learned in this topic to record your calculation.

4.4 Multiplying whole numbers by 10 and 100

We are learning to multiply whole numbers by 10 and 100.

Before we start

How many dots altogether? Write a number sentence to show how you would work out this problem.

When we multiply a number by 10, it gets 10 times bigger.
When we multiply a number by 100, it gets 100 times bigger.

Let's learn

Let's look at 3 × 10:

	HUNDREDS	TENS	ONES
3 × 10		3	3 ↙ 0

You can see the digits of the number move one place to the left, and in the space that was for the ones we put a zero. The zero is the placeholder.

What about 3 × 100?

When we multiply by 100, a number gets 100 times bigger. The digits move two places to the left, and there are two place holders.

	HUNDREDS	TENS	ONES
3 × 100	3	0	3 ↙ 0

Let's practise

1) Write the multiplication statement that you can make from each of these pictures:

a)

b)

c)

2) Write the multiplication statement to show what each of the following numbers will be when they have been multiplied by 100.

a) 7   b) 9   c) 14

d) 24   e) 56

3) There are 100 cm in a metre (m).
How many centimetres are there in each of the following:

a) 6 m   b) 9 m   c) 12 m

d) 18 m   e) 21 m

☆ CHALLENGE! ....

Amman thought of a number. He multiplied it by 10 and then by 10 again. He now has 600. What number did he start with?

What do you notice about multiplying by 10 and then by 10 again?

Work with a partner, and make up similar problems for each other.

4.5 Dividing whole numbers by 10 and 100

We are learning to divide whole numbers by 10 and 100.

Before we start

Finlay is sorting out rulers for the classroom tables. He has 20 rulers and there are five tables in the classroom. How many does he put on each table?

Write a number sentence to show this problem.

When we divide by 10, the number gets 10 times smaller.
When we divide by 100, the number gets 100 times smaller.

Let's learn

In the last unit we learned that we when we multiply by 10 the digit moves one place to the **left**, and we use zero as a placeholder. When we multiply by 100 the digits move two places to the left and there are two placeholders.

When we divide we do the opposite. To divide by 10 the digits move one space to the **right**. To divide by 100 the digits move two spaces to the right.

For example, if we shared 100 squares of chocolate between 10 people, they would get 10 each.

If we shared the 100 squares between 100 people, they would get one each.

	HUNDREDS	TENS	ONES
	1	0	0
100 ÷ 10		1	0
100 ÷ 100			1

$$100 \div 10 = 10$$
$$100 \div 100 = 1$$

1) Write a division sentence to show what each of the numbers will be when they have been divided by 10.

a) 30   b) 50   c) 20
d) 10   e) 80   f) 20

You can use cubes to help you if you need them.

2) Write division statements to show what these numbers will be when they have been divided by 10:

a) 30   b) 100   c) 50   d) 120

3) Write a multiplication and a division statement for these numbers, involving both 10 and 100. Here is an example:

1200: 12 × 100 = 1200 and 1200 ÷ 10 = 120

a) 700   b) 100   c) 900

d) 1500   e) 2400

CHALLENGE!

There are 100 centimetres in one metre. Convert these measurements from centimetres to metres.

a) 200 cm   b) 8500 cm   c) 3100 cm
d) 1400 cm   e) 2300 cm   f) 9900 cm

4.6 Solving multiplication problems by partitioning

We are learning to solve multiplication problems by partitioning tens and ones.

Before we start

Nuria has £120, made up of £10 notes. How many £10 notes does she have?

We can partition two-digit numbers into tens and ones to make it easier to solve multiplication problems.

Let's learn

Multiplying two-digit numbers by a single-digit number can be made easier by using visual representations to help partition them.

Show 12×4 using ten frames:

We can use brackets to show how we partitioned the problem:

$12 \times 4 = (10 \times 4) + (2 \times 4)$

$\qquad = 40 + 8$

$\qquad = 48$

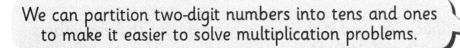

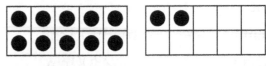

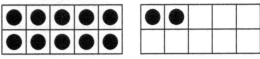

$10 \times 4 = 40 \qquad 2 \times 4 = 8$

$40 \qquad\qquad + \qquad 8 \qquad = 48$

Let's practise

1) What is 21 multiplied by 3?

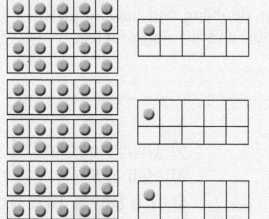

($\boxed{}$ × $\boxed{}$) + ($\boxed{}$ × $\boxed{}$)

= $\boxed{}$ + $\boxed{}$

= $\boxed{}$

2) Each class has 28 students. How many students are there in five classes?

($\boxed{}$ × $\boxed{}$) + ($\boxed{}$ × $\boxed{}$)

= $\boxed{}$ + $\boxed{}$

= $\boxed{}$

3) Isla buys four packets of sweets. There are 33 in each packet. How many sweets does she have altogether?

Use ten-frames to help you partition the problem.
Use brackets to record how you solved the problem.

★ CHALLENGE! ....

Work with a partner. Make up your own word problem that involves multiplying a two-digit number by 2, 3, 4, 5 or 10.

Swap your questions with each other and answer each other's problem. Don't forget to show your working!

4.7 Solving multiplication and division problems

We are learning to use doubling and halving facts to solve multiplication and division problems.

Before we start

True or false?
a) 20 is half of 10.
b) Double 6 is 14.
c) 8 doubled makes 16.
d) Half of 8 is 4.
e) 18 is 7 doubled.
f) 7 is half of 14.

Doubling and halving can help us solve mental calculations involving multiplication or division quickly.

Let's learn

Doubling is the same as multiplying by 2.

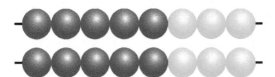

Double 8 = 16
8 × 2 = 16

Halving is the inverse of doubling. Halving is the same as dividing by 2.

Half of 16 = 8
16 ÷ 2 = 8

Let's practise

1) Double these numbers. Next, use these to make up another double.
   Here is an example: double 8 = 16, double 80 = 160
   a) Double 4   = [ ] , double [ ]   = [ ]
   b) Double 6   = [ ] , double [ ]   = [ ]
   c) Double 12 = [ ] , double [ ]   = [ ]
   d) Double 15 = [ ] , double [ ]   = [ ]

2) Answer the following problems, explaining how you worked each one out.

   a) Finlay has baked 15 samosas. Nuria has baked double that amount.
   How many samosas has Nuria baked?

   b) Isla's sister weighs 20 kg. Her brother weighs twice as much.
   How much does her brother weigh?

   c) Amman thinks there are 300 cars in a car park. Nuria thinks there are double that amount.
   How many cars does Nuria think there are in the car park?

3) Halve these numbers. When you have, use these to make up another half. Here is an example: half of 10 = 5; half of 100 = 50

   a) Half of 6 = ☐ ; half of ☐ = ☐

   b) Half of 14 = ☐ ; half of ☐ = ☐

   c) Half of 20 = ☐ ; half of ☐ = ☐

CHALLENGE!

Imagine that you win a competition with an amazing prize! You are asked to choose how you would like to receive your money:

You could be paid £100 every day for the month of June, **or ...**

You could choose to be paid £1 for the first day, but this sum would double every day for the month of June.

Which prize package would you choose? Explain why.

4.8 Recalling multiplication and division facts for 4

We are building our knowledge of multiplication and division facts for 4.

Before we start

Isla asks you to swap answers for each other to check. These are Isla's answers:

a) $2 \times 7 = 12$

b) $8 \times 2 = 18$

c) $2 \times 4 = 8$

d) Double 6 is 14

e) $20 \div 2 = 10$

f) Half of 12 is 7

Which answers are correct?

Recalling times table facts means knowing them straight away without having to think about them. This knowledge is very useful, because it helps us use strategies to solve problems more efficiently.

Let's learn

We can use our knowledge of skip counting in 2s to help us skip count in 4s. Practise skip counting in 2s first.

2, 4, 6, 8, 10, 12……

Then when you count in 4s, miss out every other number:

4, 8, 12 …

As you complete the questions, think about the patterns that you see and use these to help you recall multiplication and division facts for 4 quickly.

Let's practise

1) Use a hundred square and skip count in fours. Place a counter on all the multiples of 4.

   Can you see a pattern?

2) Use a hundred square to help you answer these questions.
   a) How many 4s are there in 36?
   b) What are five lots of 4?
   c) How many groups of 4 make 12?
   d) What are seven 4s?

3) Match the questions with the correct answers from the box.

   a) $12 \div 4$ b) $20 \div 4$

   c) 10×4 d) 4×4

   e) 7×4 f) 1×4

   g) $36 \div 4$ h) $8 \div 4$

2	40
16	3
5	28
9	4

CHALLENGE!

Finlay and his three friends make £56 at a car boot sale. How much money do they each get once this has been shared between the four of them?

Use your knowledge of multiplication facts of 4 to help you.

What other knowledge might help?

Record how you worked out the problem.

4.9 Solving division problems

We are learning to use repeated addition or subtraction to solve division problems.

Before we start

There are three cookies in each bag. Amman wants to know how many cookies he would have altogether if he had five bags.

He writes this number sentence:

$$3 + 3 + 3 + 3 + 3 = ?$$

Show him how to write a multiplication sentence to work this out.

What is the answer? Explain how you know.

We can work out division problems by adding or subtracting several times.

Let's learn

To work out 28 divided by four, we can jump back from 28 in fours and count the number of the jumps:

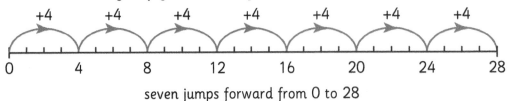

seven jumps backwards from 28 to 0

We could also jump forward in fours:

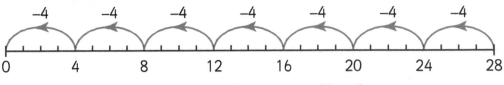

seven jumps forward from 0 to 28

There are seven jumps of four in 28, so we have worked out that $28 \div 4 = 7$.

Let's practise

1) a) Nuria has 36 beads. She makes bracelets for her friends with four beads on each one. How many bracelets can she make? Use a number line to help you.

   b) Isla has 65 seeds. She plants them in rows of five. How many rows does she plant?

   c) How many packets of nine pencils can you make from 54 pencils?

2) Amman shares 72 sweets between 12 of his friends.
   How many sweets do they have each?

3) Finlay, Nuria and Amman each took £24 on the school trip.

   Finlay spent £3 each day.

   Nuria spent £4 each day.

   Amman spent £6 each day.

   Work out how many days it will take each of them to spend all their money. Record how you worked out each problem.

   You could use a number line to help you.

CHALLENGE!

Start with 60. With which numbers can you skip count backwards and still end on zero?

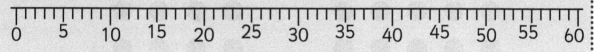

0   5   10   15   20   25   30   35   40   45   50   55   60

4.10 Multiplying by changing the order of factors

We are learning to change the order of factors to solve multiplication problems.

Before we start

Nuria's dog has made a mess of her homework! Help her work out what the numbers underneath the pawprints are.

a) $3 \times$ 🐾 $= 12$ b) $15 \div$ 🐾 $= 5$ c) $4 \times 10 =$ 🐾

d) $5 \times$ 🐾 $= 25$ e) $10 \times$ 🐾 $= 100$ f) 🐾 $\times 2 = 18$

g) $30 \div 5 =$ 🐾   h) $2 \times 7 =$ 🐾

When we are solving multiplication problems we can change the order of the factors to make the calculation easier.

Let's learn

When we multiply numbers, it doesn't matter what order the factors are in.

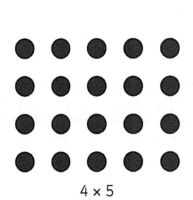

4×5

5×4

You can see the total is 20 whatever order the factors are in.

$4 \times 5 = 5 \times 4$

This is known as the **commutative property** for multiplication.

1) Use counters or cubes to make arrays for these multiplication sentences:

   a) 6 × 2   b) 3 × 4   c) 5 × 7   d) 10 × 8

   Change the order of the factors and draw an array showing what this would look like.

   Work out the total for each one.

2) Are these true or false?

   a) 2 × 9 = 9 × 2   b) 100 × 3 = 3 × 101
   c) 7 × 8 = 8 × 8   d) 4 × 15 = 15 × 4

3) Change the order of the factors for these problems:

   a) 3 × 7   b) 4 × 8
   c) 9 × 2   d) 4 × 6
   e) 5 × 9   f) 20 × 2

4) Think about each problem in question 3. Which way round would you find easier to work out?

   For each question write the factors in the order you would find easier and work out the answer.

CHALLENGE!

Work with a partner. Take a pack of playing cards and remove all the face cards. Divide the pack into roughly two equal piles and place face down.

Take turns to turn over two cards and multiply the numbers together. Discuss which order of factors you would prefer, and explain why.

For example, if I turned over 2 and 6 I could either work out 2 × 6 or 6 × 2.

4.11 Solving multiplication problems

We are learning to use facts we know and doubling to solve multiplication problems.

Before we start

What is 3 × 2?

Could you use this answer to help you work out 3 × 4?

What about 3 × 12?

We can use facts that we already know to make multiplication problems easier.

Let's learn

We can use knowledge of 2 × facts to work out multiplication problems involving 3, 4, 6 and 8 using doubling:

○ ○ ○ ○
○ ○ ○ ○

2 × 4

Let's look at 2 × 4:

To work out 3 × 4 we can add one more set of 4.

○ ○ ○ ○
○ ○ ○ ○   2 × 4   +   ● ● ● ●   1 × 4

So 3 × 4 = (2 × 4) + (1 × 4) = 8 + 4 = 12.

To work out 6 × 4, we can double the answer to 3 × 4

So 6 × 4 = (3 × 4) × 2 = 12 × 2 = 24.

To work out 12 × 4 we could use this answer and double it again.

So 12 × 4 = 24 × 2 = 48.

Let's practise

1) Make or draw an array for 2 × 7 and work out the answer.
   Use this to work out the answer to these:
   a) 3 × 7   b) 6 × 7   c) 4 × 7   d) 8 × 7   e) 12 × 7
   Record how you worked each problem out.

2) Solve these problems by first finding a known fact and then doubling.
   The first one has been done for you.

	2 ×	2 × 6 = 12	24
a)	3 ×		
b)	5 ×		
c)	4 ×		
d)	4 ×		

3) Isla went shopping. Work out how many apples she has altogether
   using facts you know and doubling.
   a) five bags of two apples   b) five bags of four apples
   c) 10 bags of four apples   d) 11 bags of four apples

CHALLENGE!

You may already know that doubling a number 4 times is the same
as multiplying by 8. For example:

4 doubled = 8   8 doubled = 16   16 doubled = 32 So, 4 × 8 is 32

a) Try this method with some other digits.

b) Now make some two-digit numbers using these
   number cards. Double your two-digit numbers
   three times in a row and then write down your
   calculation as a × 8 number fact.

4.12 Using known facts and halving to solve division problems

We are learning to use facts we know and halving to solve division problems.

Before we start

Give the answer for these 'double' questions and give the related halving fact.

a) What is 8 doubled?   b) What is 5 doubled?

c) What is 9 doubled?   d) What is 6 doubled?

We can use facts that we already know to make working out division problems easier.

Let's learn

We can use our knowledge of multiplication facts and doubling and halving to work out division problems.

half of 16 = 8

For example, if we know that double 8 is 16 then we also know that half of 16 is 8.

$8 \times 2 = 16$ so $16 \div 2 = 8$

To divide by four: half and half again.

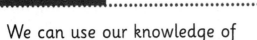

half of 16 = 8
half of 8 = 4

$16 \div 2 = 8$ so $16 \div 4 = 4$

To divide by eight: half, half and then half again.

$16 \div 8$ is half of $16 \div 4$

So $16 \div 8 = 2$.

Let's practise

1) Use these double facts to work out the missing answers. The first one has been done for you.

   a) $2 \times 8 = 16$ $16 \div 2 = 8$ $16 \div 4 = 4$
   b) $2 \times 4 = 8$ $8 \div 2 = $ __ $8 \div 4 = $ __
   c) $2 \times 6 = 12$ __ $\div 2 = $ __   __ $\div 4 = $ __
   d) $2 \times 12 = $ __   __ $\div$ __ $=$ __ $\div$ __ $=$

2) Use your knowledge of multiplication facts and halving to work out the answers to these problems:

   a) $20 \div 4$ b) $40 \div 4$ c) $40 \div 8$
   d) $36 \div 4$ e) $28 \div 4$ f) $48 \div 8$

   Record your thinking.

3) a) Finlay had 32 marbles. He shared them equally into four bags. How many marbles were in each bag? Explain your thinking.

   b) If he shared them equally into eight bags how many would there be then?

   c) Could you work out how many would be in 16 bags?

4) Nuria is planting bulbs. She has 96 altogether. If she plants them in rows of 16 how many rows will there be?

CHALLENGE!

Work with a partner. Think of a number between 1 and 10. Double it, then double it again. Take your answer and ask your partner what this would be divided by 4 to get your original number. Can they use halving to work this out?

Try doubling three times – how many would your partner have to divide by this time to get the number you started with?

What about four times?

5.1 Recognising multiples and factors

We are learning to recognise factors and multiples in multiplication facts that we know.

Before we start

Four friends pick 20 apples from the tree. They share them out equally. Write a number sentence to show how many apples each friend gets.

Knowing about factors and multiples can be very useful for developing strategies to solve multiplication and division problems.

Let's learn

Factors are numbers that divide exactly into another number.

Let's look at the number 8.

We could divide 8 by 1 to get 8.

We could divide 8 by 2 to get 4.

So the factors of 8 are 1, 2, 4 and 8.

Multiples are numbers that can be divided by another number with no remainder. Multiples are made by multiplying a number by 1, 2, 3, 4 ... and so on.

The first four multiples of 2 are 2, 4, 6 and 8.

$$2 \times 4 = 8$$

multiple of 2

multiple of 4

factor of 8   factor of 8

Let's practise

1) Identify the missing factors for these number sentences.

   a) __ × 2 = 6

   b) 10 × __ = 30

   c) __ × 3 = 12

   d) 4 × __ = 20

   Now identify any different factors for these multiples.

   e) __ × __ = 6

   f) __ × __ = 30

   g) __ × __ = 12

   h) __ × __ = 20

2) Write down the first five multiples of these numbers:

   a) 3   b) 5   c) 4   d) 10

3) Are these statements true or false?

   a) 16 is a multiple of 4.
   b) 5 is a factor of 12.
   c) 18 is not a multiple of 2.
   d) 45 is a multiple of 5.
   e) 2 and 3 are both factors of 6.
   f) 25 is a multiple of 3.
   g) 3 is a factor of 33.
   h) The first three multiples of 10 are 5, 10, 15.

CHALLENGE!

Work with a partner. Use a hundred square and put blue counters on all the multiples of 2. Now put red counters on all the multiples of 4. What do you notice?

Now put green counters on all the multiples of 3 and yellow counters on all the multiples of 5.

What patterns can you see? Does any number have all four colours of counter on it? What does this tell you?

6 Fractions, decimal fractions and percentages

6.1 Comparing fractions

We are learning to compare and order fractions.

Before we start

Using four pieces of paper, fold them to make halves, thirds, quarters and fifths.

When comparing and ordering fractions they must apply to the same-sized whole object.

Let's learn

We can make fractions by splitting an object into equal parts.

one half	one half				halves = two parts

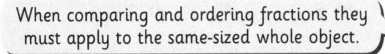

one half	one half	halves = two parts
one third	one third	one third	thirds = three parts
one quarter	one quarter	one quarter	one quarter	quarters = four parts
one fifth	one fifth	one fifth	one fifth	one fifth	fifths = five parts

We can easily compare the size of these fractions because the whole amount is the same size, so for example, we can say that:

* one quarter is larger than one fifth
* one fifth is smaller than one half.

The more parts we split the whole object into, the smaller each part will be.

We can order these fractions on a number line.

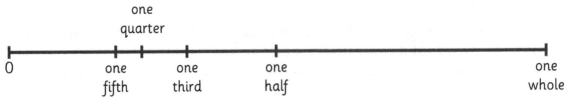

It's more difficult to compare fractions relating to different-sized bars.

one half	one half			

one third	one third	one third		

one quarter	one quarter	one quarter	one quarter	

one fifth	one fifth	one fifth	one fifth	one fifth

Here, we can't say that **one fifth of the orange bar is smaller than one half of the blue bar** because the whole objects are **different sizes.**

Let's practise

1) a) Write these fractions in the correct order from smallest to largest.

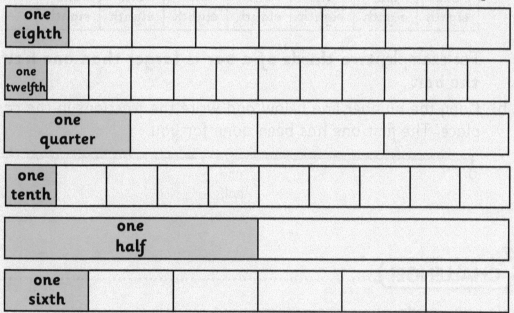

b) Copy the number line below and write the fractions in the correct place. The first one has been done for you.

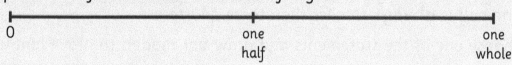

0   one half   one whole

2) a) Use the bar models below to write three statements that compare one fraction to another.

one half	

one third	one third	

one quarter	one quarter	one quarter	

one fifth	one fifth			

one sixth	one sixth			

one eighth	one eighth	one eighth	one eighth	one eighth	one eighth	one eighth	

For example, **two thirds of a bar is larger than one half of the bar.**

b) Copy the number line below and write the fractions in the correct place. The first one has been done for you.

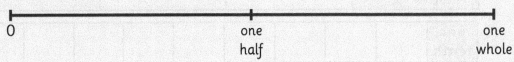

0   one half   one whole

CHALLENGE!

Finlay has made some statements about fractions:

One half is **always** larger than one quarter.
One third is **always** larger than one fifth.
One half is **always** smaller than three quarters.

Choose one of the statements and draw bar models to prove him wrong.

6 Fractions, decimal fractions and percentages

6.2 Identifying equivalent fractions pictorially

We are learning to identify equivalent fractions visually.

Before we start

Sometimes, always, never?
One half is always equal to two quarters.
Draw pictures to justify your answer.

'Equivalent' means 'equal'. The size of the 'portion' remains exactly the same but the number of parts that make it up changes because the size of the parts changes.

Let's learn

If we get half a pizza, we can cut it into two equal parts and get one of those parts, one half. We can cut the same pizza into four equal parts and get two of those parts, two quarters. The amount of pizza we get is exactly the same. The parts are smaller so we get more parts in order to get the same amount.

Equivalence is about splitting an object or amount in different ways without changing the size or amount of the portion.

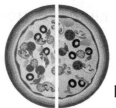

is the same sized portion as

It is important when working out equivalent fractions that we are talking about the **same-sized object** (or amount). One half of the pizza on the left is **not** the same as two quarters of the one on the right

is NOT the same sized portion as

because the two pizzas are different sizes in the first place.

6

1) Amman wants to eat half a pizza.
Which portion can he choose so that he gets
exactly half? Name the portions.

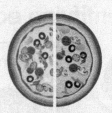 one half

a)

b)

c)

d)

e)

f)

g)

h)

i)

2) Nuria is to get one third of a pizza.

   a) Identify and name the portions that are equivalent to one third.
   b) Make up other portions that are equivalent to one third.

3) Isla is to get three quarters of a pizza.

   a) Identify and name the portions that are equivalent to
      three quarters.
   b) Make up other portions that are equivalent to three quarters.

CHALLENGE!

Using the fraction wall below, can you identify sets of fractions that are equivalent? How many different sets can you find?

one whole											
one half						one half					
one third				one third				one third			
one quarter			one quarter			one quarter			one quarter		
one fifth		one fifth		one fifth		one fifth			one fifth		
one sixth		one sixth		one sixth		one sixth		one sixth		one sixth	
one eighth	one eighth	one eighth	one eighth	one eighth	one eighth	one eighth	one eighth				
one tenth	one tenth	one tenth	one tenth	one tenth	one tenth	one tenth	one tenth	one tenth	one tenth		
one twelfth	one twelfth	one twelfth	one twelfth	one twelfth	one twelfth	one twelfth	one twelfth	one twelfth	one twelfth	one twelfth	one twelfth

6 Fractions, decimal fractions and percentages

6.3 Identifying and creating equivalent fractions

We are learning to identify and create equivalent fractions.

Before we start

Name three fractions that are equal to one half. Explain how you would work this out.

Equivalence has nothing to do with multiplying or dividing a fraction!

Let's learn

When we find equivalent fractions the size of the portion remains exactly the same. What we do is split an object, or amount, into smaller or bigger parts until we find another fraction of the same size. The number of fractions that are equivalent is never ending.

Halves split into two equal parts creates quarters. One half = two quarters.

Halves split into three equal parts creates sixths. One half = three sixths.

Halves split into four equal parts creates eighths. One half = four eighths.

| one half | two quarters | three sixths | four eighths |

Let's practise

1) Finlay wants to find out what quarters change into when he splits them into:

   a) two parts   b) three parts   c) five parts

   Help him by naming each of these fractions and writing fractions that are equal to one quarter.

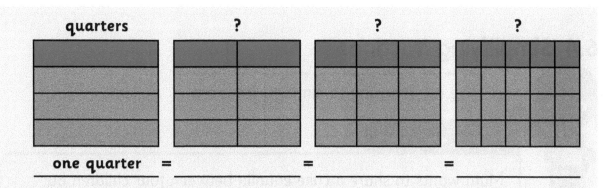

quarters   ?   ?   ?

one quarter = _____ = _____ = _____

2) Write three other fractions that are equivalent to three quarters.

3) Isla wonders if quarters can be changed into hundredths.

   a) How many parts would each quarter have to be split into?

   b) Write equivalents for the following.

      i) One quarter = ____ hundredths

      ii) Two quarters = ____ hundredths

      iii) Three quarters = ____ hundredths

      iv) Four quarters = ____ hundredths

CHALLENGE!

a) Using equal-sized pieces of paper (for example, post-it notes), fold them to make halves, thirds, quarters and fifths. Fold these to find equivalent fractions. Find fractions that are equivalent to:

   a) one third   b) three quarters

   c) two thirds   d) three fifths

b) How many equivalent fractions can you find for each of the above?

c) How many different sets of equivalent fractions can you find?

Fractions, decimal fractions and percentages

6.4 Simplifying fractions

We are learning to simplify fractions.

Before we start

Mum wants to share a cake equally between four children but they've said they don't want her to cut it into quarters. What else could she do so that they get exactly the same amount each?

Simplifying a fraction is about finding an equivalent fraction that is simpler to make.

Let's learn

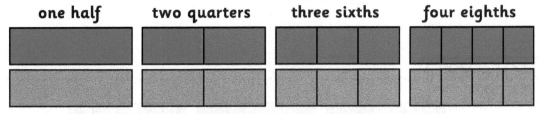

Equivalent fractions are fractions that are equal in size.

We have learned that **one half = two quarters = three sixths = four eighths** and so on.

one half	two quarters	three sixths	four eighths

When we *simplify* a fraction we find an equivalent fraction that is *simpler* to make. For example:

Finlay cuts a block of modelling clay into 10 equal parts (tenths) to share with Isla.

They take five tenths each.

Isla says: 'There is a simpler way to do that! You could have cut the block into two equal parts and we could have taken one half each.'

Isla has suggested a way of making this task **simpler** for Finlay.

We can see that **five tenths** can be simplified to make **one half**.

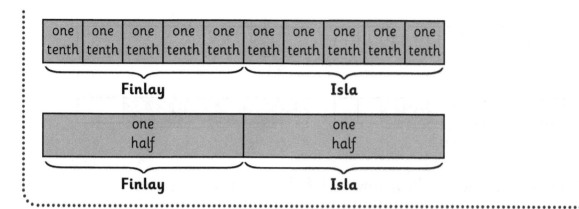

Let's practise

1) Match each green fraction to the correct blue simplified version.

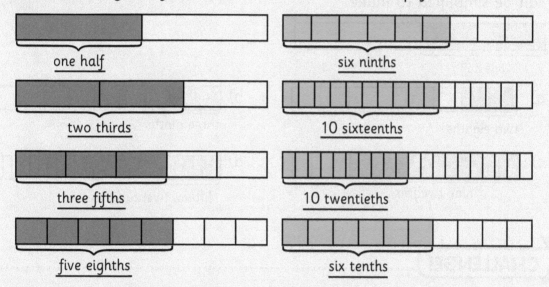

one half   six ninths

two thirds   10 sixteenths

three fifths   10 twentieths

five eighths   six tenths

2) Identify how these fractions have been simplified.

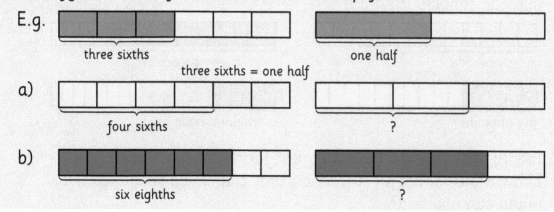

E.g.

three sixths   one half

three sixths = one half

a) four sixths   ?

b) six eighths   ?

c)

five twentieths

?

d)

eight tenths

?

3) Draw bar models to show how the following fractions can be simplified.

two quarters

can be simplified to make:

one half

a)

two eighths

b)

three ninths

c)

nine twelfths

d)

fifteen twenty-fifths

CHALLENGE!

a) Choose one of the fractions below and see how many different ways you can simplify it.

six twelfths

twelve sixteenths

five fifteenths

twelve twentieths

b) Identify the fraction in its simplest form.

c) Draw a bar model for a fraction that cannot be simplified. How many can you find?

6.5 Writing decimal equivalents to tenths

We are learning to find the decimal equivalent of a fraction (tenths).

Before we start

Finlay has been given the two fraction cards below and has been asked to convert them into tenths.

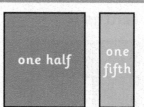

one half | one fifth

Finlay says: 'If I want to convert these into tenths, I just need to cut them both into ten equal pieces.'

Do you agree with Finlay? What advice would you give him?

Fractions can be represented in different ways. They can have the same value but appear differently.

Let's learn

The number system we use is called the **decimal system**. 'Deci' means ten. This is because each place in a number gets ten times larger or smaller.

The first place after the decimal point is tenths (because one tenth is 10 times smaller than one).

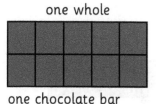

one whole

one chocolate bar

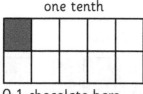

one tenth

0·1 chocolate bars

One tenth is ten times smaller than one whole. We can write this as 0·1.

one tenth = 0·1

1) Write the amount of chocolate for each diagram as a fraction and a decimal fraction.

a)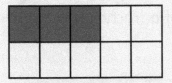

b)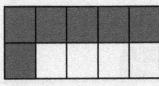

three tenths = _____

_____ = _____

c)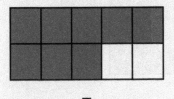

d)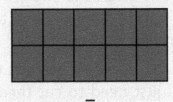

_____ = _____

_____ = _____

2) Finlay has half a bar of chocolate. He wants to be able to write this as a decimal fraction. Help him.

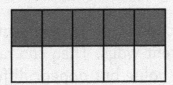

one half = _____ = _____

3) Isla has two fifths of a cake. Help her convert this into a decimal fraction.

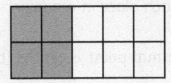

two fifths = _____ = _____

4) Amman has this amount of chocolate.

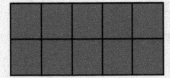 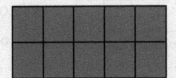 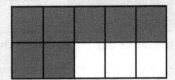

He can write this as:
two whole and seven tenths or **27 tenths** or **2·7**

a) Write the following amounts in three different ways.

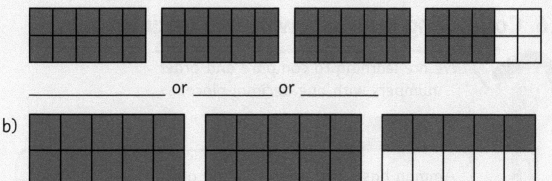

_____ or _____ or _____

b)

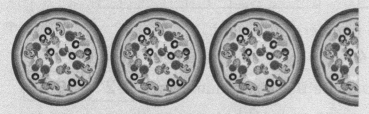

_____ or _____ or _____

⭐ **CHALLENGE!** ....

Amman is finding it difficult to write the portions below as decimal fractions.

Three and a half pizzas:

Two and four fifths cakes:

Amman says: 'I can't write these as decimal fractions because there are no tenths.'

Isla says: 'Can you use equivalent fractions to help you convert them into tenths?'

a) Help Amman write each portion as a decimal fraction.

b) Make up three more examples you could use to help Amman understand.

6 Fractions, decimal fractions and percentages

6.6 Comparing numbers with one decimal place

We are learning to compare and order numbers with one decimal place.

Before we start

Amman has been asked to write a list of five decimal numbers that are larger than 9 but smaller than 10 in order from smallest to largest. Help him write the list.

Decimal fractions help us compare numbers with greater accuracy.

Let's learn

0·1 = one tenth

0·4 = four tenths

We can compare these two numbers by saying:

0·1 is smaller than 0·4   or   **0·4 is larger than 0·1**

Finlay has been asked to compare these two numbers: | 1·2 | | 0·9 |

Amman says: '1·2 has two tenths and 0·9 has nine tenths, so I can say **0·9 is larger than 1·2.**'

Nuria says: 'I don't think that's right. Let's draw a bar model for each number to check.'

| 1·2 |

| 0·9 |

1·2 is actually one whole bar plus two tenths, so what Finlay should have said is:

'1·2 is larger than 0·9.'

1·2 is the same as 12 tenths which is larger than nine tenths.

Let's practise

1) Write a statement using decimal numbers to compare each of the bar models below. The first one has been done for you.

a)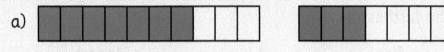

0·7 is bigger than 0·3

b)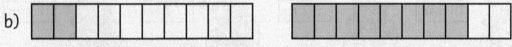

c)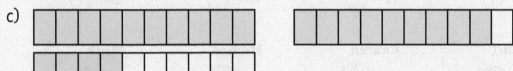

d)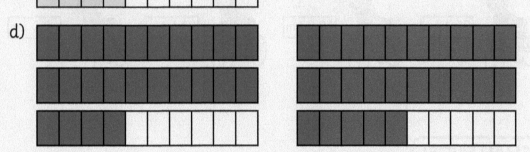

2) Copy the number line below and write the numbers in the correct positions. The first one has been done for you.

Make up four statements using these numbers. For example, **0·4 is smaller than 0·8.**

| 0·8 | 2·2 | 1·8 | 1·0 | 1·7 | 1·5 | 0·5 | 2·0 | 0·4 | 1·4 |

3) The cyclists below are racing from Glasgow to Edinburgh. Their screens show the distance they have travelled so far. Write all the names in the correct order from the cyclist who has travelled the least distance to the furthest.

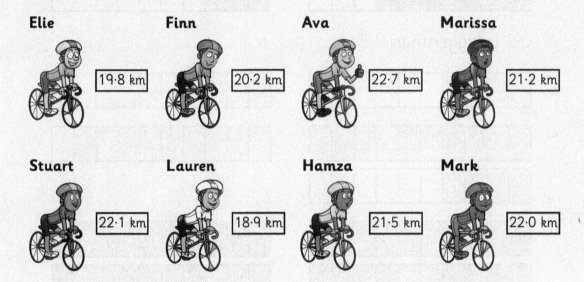

Elie 19·8 km **Finn** 20·2 km **Ava** 22·7 km **Marissa** 21·2 km

Stuart 22·1 km **Lauren** 18·9 km **Hamza** 21·5 km **Mark** 22·0 km

⭐ **CHALLENGE!**

a) Finlay has been given the following pairs of number cards and has been asked to write the numbers that lie in between.

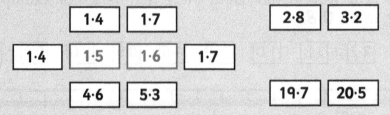

| 1·4 | 1·7 | | 2·8 | 3·2 |

| 1·4 | 1·5 | 1·6 | 1·7 |

| 4·6 | 5·3 | | 19·7 | 20·5 |

He's had a go at the first pair of cards.

Help Finlay write the numbers in between for each pair.

b) Nuria has been given these number cards. **0·4** **0·5**

Amman says he can think of nine numbers that lie in between the two numbers.

Find some or all of the numbers that Amman is thinking about. Could Finlay have found any more numbers?

6.7 Calculating a simple fraction of a value

We are learning to solve problems by calculating a fraction of a value.

Before we start

Nuria has 24 sweets. She's been asked to give one third of them to Finlay but she doesn't understand how to work this out. Explain to Nuria how she could solve this problem and find out how many sweets to give to Finlay.

Bar models can help us work out a fraction of an amount.

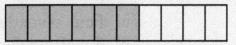

Let's learn

Finn spends *five sevenths* of his money on a new cap. How much was the cap if he had **£21** to begin with?

Sevenths tells us how many parts to split the money into:

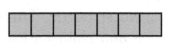

Five tells us how many parts relate to the cap:

starts with

cap   left over

£21 tells us the whole amount to be shared out

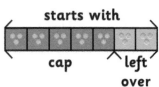

starts with

cap   left over

£21 shared between seven parts = £3 in each part.

Five parts out of seven was spent on the cap so we only count the money in these parts: 5 × £3 = £15, so the cap cost £15.

1) Use the bar models to work out the following.

   a) Two thirds of 18.

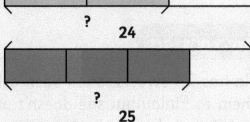

   b) Three quarters of 24.

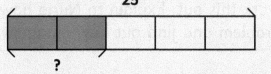

   c) Two fifths of 25.

2) Draw a bar model to show how you could work out the following.

   a) Five sixths of 36
   c) Three eighths of 32

   b) Four sevenths of 35
   d) Seven tenths of 30

3) Draw a bar model to solve each of the following problems.

   a) Five eighths of the children in a class are girls. How many girls are in the class if there are 32 children altogether?

   b) Seven tenths of the children at football training have brought a water bottle. There are 60 children at training altogether. How many children have brought a water bottle?

   c) Finlay spent six ninths of his birthday money on a new pair of football boots. He got £63 altogether for his birthday. How much did he spend on the boots?

CHALLENGE!

Make up and solve word problems for any of the following bar models.

a)
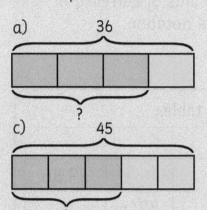
36

?

b)
28

?

c)
45

?

d)
40

?

Three thirds of the sweets in a packet are lemon-flavoured. There are 36 sweets altogether. How many lemon sweets are there?

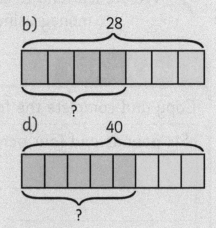

Nuria has attempted a question for the first bar model.

Does her problem match the bar model?

Can you give her any advice?

7.1 Writing amounts using decimal notation

We are learning to write amounts of currency or money using decimal notation.

Before we start

Copy and complete the following table:

Six pounds and four pence		
		£9.36
	824p	
Nine pounds and forty pence		
	23p	

We can write amounts of money in different ways but the value of the amount stays the same.

Let's learn

The notes and coins that we use can be recorded in a variety of different ways, but the value will always be the same.

£4·40 or 440p or four pounds and forty pence

Let's practise

1) Write the following amounts using decimal notation:
   a) 523p   b) 94p   c) Twelve pounds and eight pence

For the following questions write the total amount using decimal notation:

d) Six £5 notes

e) One hundred one pence pieces

f) 12 five pence pieces

2) Meg the Baker sells rolls for 45p each.

Mr Todd bought a dozen rolls. He had six £1 coins.

a) Did he have enough money?

b) How much change did he get?

3) During the holidays Isla was given 50p every day from her papa and £1 from her mum.

She wanted to buy a new pair of jeans that cost £19·50.

How many days will it take her to save up for her jeans?

⭐ CHALLENGE! ....

Investigate with a partner two different currencies used in other countries in the world.

For each currency, record the following information in any way you choose (use either words, pictures or diagrams):

• Name of the country

• Name of the currency

• Notes used in the currency

• Coins used in the currency

Money

7.2 Budgeting

We are learning to manage and keep to a budget.

Before we start

Finlay went to the shops and his receipt is shown here.

Calculate the total amount he spent and the change he would have from £20.

```
              RECEIPT
Tex#: 5476      05-02-2018 11:32 AM

   Item      Qty      Rate

Crisps       1       £2·50
Juice        1       £1·65
DVD          1       £7·99
Popcorn   1       £2·99
************************
  TOTAL AMOUNT
--------- THANK YOU ---------
```

When we budget our money we can decide if we can afford to buy something.

Let's learn

A **budget** is the amount of money that someone has to spend.

Sometimes we have to change the quality of items we want to buy, or the quantity, depending on what our budget is.

If we have less money than we need, we can remove one or more items to make the purchases affordable or we can wait until we have more money in our budget.

If we have more money than we need to pay for a set of items, we can add one or more items to spend the full amount.

Amman saves £40 a week so he can buy a new guitar that costs £250.

If he saves for six weeks, will he be able to afford his new guitar?

£40 × 6 = £240. Therefore he will **not** have enough.

He will need another £10.

Let's practise

1) Finlay wants to buy a new pair of football boots that cost £125.

   He has already saved £72 from his birthday money.

   He gets £5 per week pocket money.

   For how many weeks will he need to save his pocket money before he can afford to buy his boots?

2) Nuria would like to buy a new outfit for a party.
   Her total budget is £50.

Vest top	Skirt	Crop jeans	T-shirt	Hat and bag
£15·50	£32·50	£29·50	£18·25	£14·50

   Look at the items below and work out which of the following outfits she can afford.

   a) Skirt and T-shirt
   b) Skirt, vest top, hat and bag
   c) Crop jeans and vest top

3) Nuria works five days a week and has budgeted £25 a week to pay for her journey to and from work.

   The bus fare is £2·20 for a single ticket.

   The train fare is £4·25 return per day.

   a) Which mode of transport should Nuria take?
   b) How much will her total fare be per week?
   c) How much change will she receive?

CHALLENGE!

Isla is planning a trip to the cinema for her birthday.

Her parents have given her a total budget of £75 and it is up to her how many friends she can take and the snacks they can have.

If she wants to buy everyone a ticket, large popcorn, large drink and a three-scoop ice cream:

a) How many friends can she take to the cinema?

b) How much will it cost?

c) Will she have any budget left over?

Child ticket	£8·50 each
Group ticket	£30·00 for four

Small popcorn	£2·50
Medium popcorn	£3·25
Large popcorn	£4·00

Small drink	£2·25
Medium drink	£3·75
Large drink	£3·15

one scoop	£1·90
two scoops	£2·40
three scoops	£2·85

7.3 Saving money

We are learning about saving money by comparing costs.

Before we start

Nuria has £10 to spend.

She must buy bread, milk, jam, tea bags and sugar for breakfast.

What is the **least amount** she will need to spend?

Baguettes three for £2

Sliced loaf £1·50

Carton of milk 85p

Jam £2·90

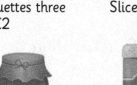

Tea £1·80

Bag of sugar £2·50
Sugar cubes £2 per box

You can save money by comparing the cost of items you are buying and then choosing the cheapest option.

Let's learn

When we are buying something there are often different prices for the same product. This depends on how many you buy, which shop you buy it from and how you choose to pay for it.

 five for £2·50

 eight for £3·60

To ensure that you get the best value for money, you need to **compare** the different offers.

Which option is the better buy?

Pack of five cookies at £2·50 → 250p divided by 5 = **50p** each
Pack of eight cookies at £3·60 → 360p divided by 8 = **45p** each

Therefore, the larger packet is cheaper by 5p per cookie.

Let's practise

1) Isla needs 10 cooking apples to make a pie.

How much will she save by buying the crate of 10 rather than 10 individual apples?

 38p each

 Crate of 10 for £3·60

2) Nuria's mum needs a ticket for her daily journey to and from work.

She works five days per week, so she needs 10 tickets per week.

Forthside Bus Company charges £1·20 per ticket or £10·50 for a book of 10 tickets.

Central Bus Company charges £2·20 for a return ticket.

N-Line Coaches charges £55 for 50 tickets.

a) What is the cheapest ticket Nuria's mum can buy and from which company?

b) How much per week will she save if she buys a book of 10 tickets from Forthside Bus Company over tickets from Central Bus Company?

3) The Roberts family want to book a holiday for the best possible price.

Mr Roberts, Ms Roberts, their sons John (6) and Christopher (8) and their daughter Millie (14) all want to spend a week at a villa in Spain. They want to have a car for all 7 days.

Forever Travel are advertising seven nights accommodation for £250 per person but flights are not included.

Roland Air have return flights for £95 per person with optional car hire for an extra £10 per day.

Vibrant Villas are advertising seven nights accommodation including return flights and car hire for £375 per person.

a) What is the least amount the Robertson family will pay?
b) Who will they be cheaper travelling with?
c) How much will they save?

CHALLENGE!

Finlay's aunt is trying to save money on her weekly shopping, so for the last three weeks she has been to different shops and bought the same items.

Sainsbiddies Supermarket

five litres of rice milk @ 35p/litre

two boxes of porridge @ £2/box

four litres of sparkling water @ 40p/litre

four tins of chopped tomatoes @ 25p/tin

700 g of turkey mince @ 25p/100 g

Total

Murry's Marketplace

five litres of rice milk @ 32p/litre

two boxes of porridge @ £2·10/box

four litres of sparkling water @ 38p/litre

four tins of chopped tomatoes @ 15p/tin

700 g of turkey mince @ 27p/100 g

Total

Supersavers Superstore

five litres of rice milk @ 30p/litre

two boxes of porridge @ £1·90/box

four litres of sparkling water @ 45p/litre

four tins of chopped tomatoes @ 20p/tin

700 g of turkey mince @ 27p/100 g

Total

a) Which shop was **cheapest**?
b) How much did Finlay's aunt **save** compared to the most expensive shop?
c) If she visits all three shops to buy the cheapest items, what is the **least** she will spend?

7.4 Profit and loss

We are learning to calculate basic profit or loss.

Before we start

Amman spent his birthday money on the following items:

Put the items in order from what you think is the least expensive to the most expensive. Check with a partner that you agree on the order.

A profit or loss is the difference in the amount of the selling price and the buying price.

Let's learn

A **profit** is the amount of money that is made when something is sold for **more than** it cost to buy it.

The phone cost Isla £55 and she sold it for £70.

She made a £15 **profit** (£70 − £55 = £15).

A **loss** is the amount of money that is lost when something is sold for **less than** it cost to buy it.

Nuria bought a bag for £100 and sold it for £65.

She made a **loss** of £35 (£100 − £65 = £35).

Let's practise

1) Isla bought a signed football strip for £85 at an auction.
She sold it online for £175.
How much of a **profit** did she make?

2) Finlay bought a collection of four paintings for a total of £1250.
He sold each one for £425.
How much of a **profit** did he make?

3) Amman bought two tickets for a concert to see his favourite
band for £75 each. Unfortunately, he could not go so
he sold both tickets for £120 in total.
Did he make a **profit** or a **loss** and by how much?

4) Nuria's mum bought a classic car for £7500 about five years ago.
She recently sold it for a **profit** of £2375.
How much did she sell it for?

CHALLENGE!

Isla's sister makes dolls houses.
The basic house kit costs £25.
The deluxe house kit costs £35.
The materials to decorate the house cost £14.
The basic furniture pack costs £25.
The deluxe furniture pack costs £32.
She charges £10 per hour for her time and it takes her eight hours
to make each house.
Isla's sister likes to make a £20 profit on a basic house and £30 on
a deluxe house.

a) How much does the basic dolls house cost her customer?

b) How much does the deluxe dolls house cost her customer?

8.1 Telling the time to the minute – 12-hour clock

We are learning to read and write 12-hour time to the nearest minute.

Before we start

Copy and complete the table filling in the missing information:

		Fifteen minutes past three
	12:35	

Time can be recorded in different ways, including analogue, digital, 12-h and 24-h.

Let's learn

A digital clock has only numbers. An analogue clock has hands that rotate over a numbered scale (normally 1–12).

Unlike analogue clocks, digital clocks normally display whether the time is in the morning or afternoon/evening.

A 12-hour digital clock uses am and pm to indicate morning and afternoon/evening.

am is the time from 12 midnight and before 12 noon.

pm is the time from 12 noon and before 12 midnight.

11:13 AM

Let's practise

1) Write the time shown in words. The first one has been done for you.

a) **5:01**
One minute past five

b) **3:54**

c) **12:22**

d) **8:57**

e)

f)

g)

h)

2) Draw the hands on an empty clock face to show the times given.

a) Twenty-one minutes past one

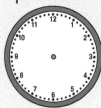

b) Twenty-nine minutes to nine

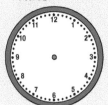

c) Six minutes past 12

d) Nine minutes to 11

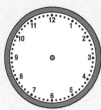

3) Write the time shown. Remember to include the time of day: morning, afternoon or evening.

a) **8:44** AM   b) **4:18** AM   c) **7:49** PM   d) **2:23** PM

4) Convert the times from analogue to digital, or from digital to analogue. Then write the time shown.

a)

		PM

b) | 11 | 19 | AM |

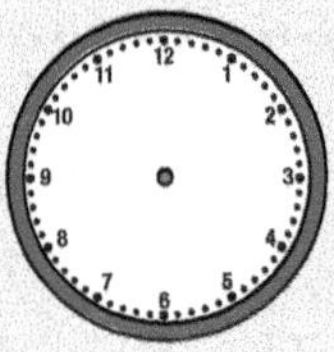

c)

		AM

d) | 9 | 57 | AM |

CHALLENGE!

Work with a partner, taking turns to throw a dice and move along the track.

If you land on an analogue clock, read the time shown, then write it.

If you land on a digital clock, read the time shown, write it and model it on an analogue clock.

If you read and write the time correctly, you get an extra turn.

The first player to reach 'FINISH' is the winner.

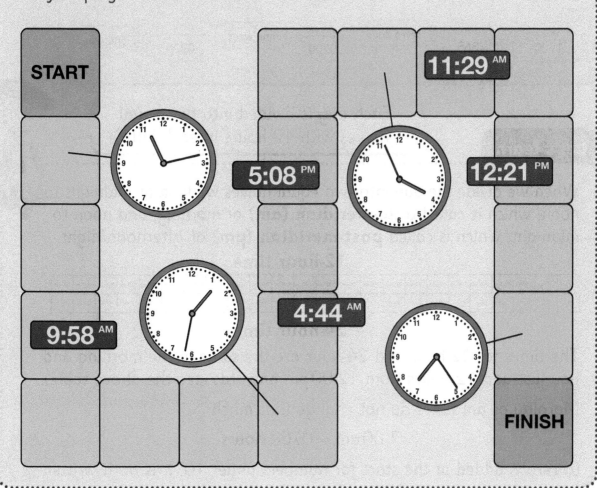

START

11:29 AM

5:08 PM

12:21 PM

4:44 AM

9:58 AM

FINISH

8.2 Converting between 12-h and 24-h time

We are learning to change 12-h time to 24-h time.

Before we start

Match the following time facts:

| 24 hours | 2:00 am | 4:00 am | | 5:00 pm | one year | 1600 |

| 3:00 am |

| 1700 | 365 days | seven days | 0200 | one week | 0300 | one day |

Each day is divided into two equal halves with 12 hours in each half.

Let's learn

When we divide the day into two equal halves we look at midnight to noon, which is called **ante-meridian (am)** or morning, and noon to midnight, which is called **post-meridian (pm)** or afternoon/night.

12-hour time

12:00 am	1:00 am	2:00 am	3:00 am	4:00 am	5:00 am	6:00 am	7:00 am	8:00 am	9:00 am	10:00 am	11:00 am	12:00 pm	1:00 pm	2:00 pm	3:00 pm	4:00 pm	5:00 pm	6:00 pm	7:00 pm	8:00 pm	9:00 pm	10:00 pm	11:00 pm
0000	0100	0200	0300	0400	0500	0600	0700	0800	0900	1000	1100	1200	1300	1400	1500	1600	1700	1800	1900	2000	2100	2200	2300

24-hour time

The times for 12-hour and 24-hour are the same in the morning and you just add 12 hours from 12:00 pm onwards (see the above table).

Morning or am times do not change that much:

7:00 am = 0700 hours

(a zero is added at the start for numbers under 10, it is both written and said – 'zero seven hundred hours')

11:33 am = 1133 hours

However, afternoon, evening or pm times will have 12 hours added:

4:00 pm = 1600 hours (12 o'clock + 4 hours)

9:10 pm = 2110 hours (12 + 9)

Let's practise

1) Convert the following 12-hour times to 24-hour times:
   a) 9:35 pm
   b) 3:30 am
   c) 4:45 am

2) Convert the following 24-hour times to 12-hour times:
   a) 1840
   b) 0340
   c) 2122

3) Write each of these times in 12-hour and 24-hour times:
   a) **8:44** AM
   b) **7:49** PM

   c) afternoon
   d) morning

CHALLENGE!

Copy and complete the following table filling in the missing times:

12-hour time	24-hour time
2:15 am	
6:46 pm	
	2104
Noon	
	0428
	1314

8.3 Converting minute intervals to fractions of an hour

We are learning to convert minute intervals to fractions of an hour.

Before we start

On her way to school it takes Isla $\frac{1}{4}$ of an hour to walk to the bus stop, her bus journey lasts $\frac{1}{2}$ an hour and then she walks for $\frac{1}{4}$ of an hour before she arrives at school.
How many minutes does her total journey to school take?

Every hour can be divided equally into two halves or four quarters.

Let's learn

There are 60 minutes in every hour.

An hour or 60 minutes can be divided equally into two halves of 30 minutes.

    2:30 pm = half past two in the afternoon.

An hour or 60 minutes can also be divided equally into four quarters of 15 minutes.

    6:15 am = quarter past six in the morning.

Let's practise

1) Draw the hands on an empty clock face to represent the times written above each one:

   a) Quarter past six   b) Half past nine

   c) Quarter to eleven   d) One o'clock

2) Write the times shown on each clock using words. Remember to use 'quarter to', 'quarter past' or 'half past' as appropriate.

a)    b)   c)    d)

CHALLENGE!

1) Record your answers to the questions below in three ways – digital, analogue and word form:

a) Amman was going to get the bus in half an hour.

If it is quarter past nine in the morning now, what time does his bus leave?

b) Isla has an appointment at 12:45 pm.

The appointment will last for three quarters of an hour.

What time will her appointment finish?

c) Finlay went for his break 15 minutes ago.

He has half an hour before his break is over when it will be 11 am.

What time did his break begin?

2) Amman's dad usually finishes work at midday on a Friday.

This week he wants to finish at 10 am instead to leave for his holiday.

If he works an extra half an hour on Monday, and quarter of an hour extra on both Tuesday and Wednesday, how much longer will he have to work on Thursday so that he has worked enough to finish at 10 am on Friday?

8 Time

8.4 Calculating time intervals or durations

We are learning to calculate time intervals.

Before we start

Amman is watching a film. He starts watching it at 7 pm and it finishes at 8:35 pm.

How long does the film last?

A time interval or duration can be calculated using seconds, minutes, hours or days.

Let's learn

One way to calculate how long something lasts is by splitting up the time into chunks or 'chunking' it.

If a bus starts its journey at 6:30 am and reaches its destination at 8:45 am, how long did the journey last altogether?

| one hour/60 minutes | + | one hour/60 minutes | + | 15 minutes | = two hours 15 minutes |

6:30 am   7:30 am   8:30 am   8:45 am

Let's practise

1) Copy and complete the timeline and work out the total time interval from start to finish:

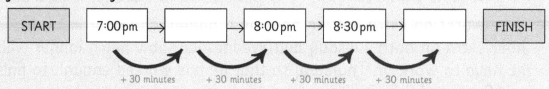

| START | 7:00 pm | | 8:00 pm | 8:30 pm | | FINISH |

+ 30 minutes   + 30 minutes   + 30 minutes   + 30 minutes

2) Calculate the following time intervals using the 'chunking' method:

a) Midnight to 1:30 am

b) 11:30 am to 2:00 pm

c) 1920 to 2100

d) 9:00 pm to 11:00 pm

e) Midday to 3:00 pm

f) 2200 to 0230

3) Using the information from the train timetable below, answer the following questions:

Edinburgh	0830	0900	0930
Haymarket	0840	0910	0940
Edinburgh Park	0850	0920	0950
Linlithgow	0900	0930	1000
Polmont	0910	0940	1010
Falkirk Grahamston	0920	0950	1020
Camelon	0930	1000	1030

I arrive at Edinburgh train station at 8:20 am to get a train to Linlithgow.

a) How long do I need to wait until the next train leaves?

b) How long does the Haymarket train take to arrive at Camelon?

c) It takes Isla 10 minutes to walk from her house to the train station.

If she leaves the house at 8:30 am, when is the next train she can get from Linlithgow?

CHALLENGE!

The McBride family are going to New York on holiday.

Their plane leaves Glasgow airport at 2345 on Saturday.

The duration of the flight is seven hours and 35 minutes.

New York time is five hours behind UK time.

What time will it be in New York when the McBride family reach their destination?

8.5 Speed, time and distance calculations

> We are learning to calculate speed, time and distance.

Before we start

> Nuria is looking at the bus timetable and she notices that the buses are every 30 minutes.
>
> If the first bus leaves the station at 7:40 am, at what time will the fourth bus leave the station?

> Three factors are taken into consideration when calculating a journey: **distance**, **speed** and **time**.

Let's learn

To calculate the distance that you have travelled you can use the following formula:

$$distance = speed \times time$$

An easy way to remember this is D Equals S T or **DEST**, which is the first part of the word 'destination'.

If a car is travelling at 50 km per hour for four hours, how far will it go?

Using the formula and the information in the question, we can complete the following calculation:

$$D = S \times T$$
$$D = 50 \times 4$$
$$D = \textbf{200 km}$$

Let's practise

1) Use the formula distance = speed × time to calculate the distance travelled by the following people:

   a) If Amman was driving at 45 mph for three hours, how far did he drive?

   b) If Isla walked at 5 mph for four hours, how far did she walk in total?

2) Use the formula **D = S × T** to calculate the distance travelled by each of these modes of transport:

   a) A plane was in the air for six hours, travelling at a speed of 360 mph. What was the total distance it travelled?

   b) The overnight train travelled at 120 mph for seven hours before it reached its destination. How far did it go?

3) A bus left Stirling at 10:35 am and arrived in Aberdeen at 1:35 pm. It travelled at 65 mph.

   How long did the journey take and how many miles did the bus travel altogether?

CHALLENGE!

Mr and Ms Tortolano decided to drive from their home in Forres to their daughter's house in Galashiels.

They left at 7:30 am and stopped in Perth at 9:30 am for breakfast.

After a half-hour break they continued their journey and stopped in Edinburgh for lunch at midday.

One hour later they started the final leg of their journey and arrived at their destination at 2 pm.

On average they drove at 50 mph on the first and third parts of their journey and 70 mph on the second.

What was the total distance they drove?

9.1 Using familiar objects to estimate length, mass, area and capacity

> We are learning to estimate measurements by comparing familiar objects.

Before we start

What are the units of measurement for each of the items below?

Length:

Area:

Mass:

Capacity:

We can compare objects with those we know, to estimate their measurements.

Let's learn

If we know the measurements of something, we can use these to estimate the measurements of other things.

Nuria says: 'The pencil looks about double the length of the key. I estimate that the pencil is 10 cm long.'

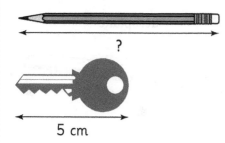

Isla says: 'The water bottle feels half as heavy as the lunchbox. I estimate that it weighs 100 g.'

Amman says: 'The bottle holds about three times as much water as the glass. I think it holds about 300 ml.'

Finlay says: 'The big square looks about four times bigger than the little square. I think its area is about four square metres.'

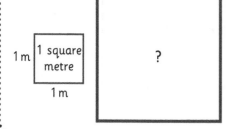

Let's practise

1) The length of the bike is 2 m. Compare the length of the bike with the other items and estimate their lengths.

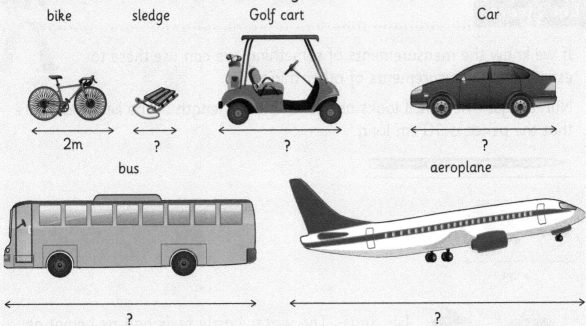

bike   sledge   Golf cart   Car

2m   ?   ?   ?

bus   aeroplane

?   ?

2) The area of the cupboard is four square metres (4 m²). Estimate the area of the other rooms in the house.

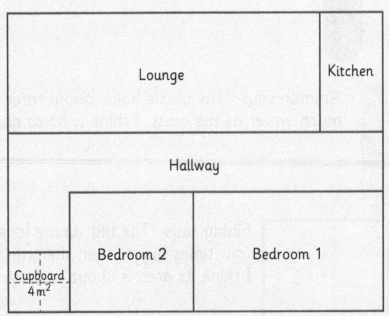

Lounge

Kitchen

Hallway

Bedroom 2

Bedroom 1

Cupboard
4 m²

CHALLENGE! ....

a) You will need:
 • a one kilogram weight
 • Five items from your environment.
Copy the table below.

Item	Estimated Mass	Actual Mass

Hold the one kilogram weight in one hand and one of the items in the other hand.

Compare the weight of them both to estimate the weight.

Measure the weight using suitable scales.

b) You will need:
 • a plastic cup
 • a funnel
 • five different-sized plastic bottles (numbered 1 to 5).
Copy the table below.

Estimate how many cups of water it will take to fill each of the bottles.

Use the funnel to find out how many cupfuls each bottle holds.

Bottle	Estimated Capacity	Actual Capacity
1		
2		
3		
4		
5		

9 Measurement

9.2 Estimating and measuring length

> We are learning to estimate and measure the length of an object.

Before we start

Measure the length and height of your table in hands.

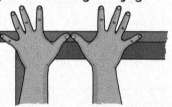

> The **length** of an object is the distance from one end to the other.

Let's learn

We can measure length using:

ruler	metre stick	tape measure	trundle wheel

We use **metres** to measure longer lengths.

1m 1m 1m 1m 1m 1m

The length of this car is about six metres.

We use **centimetres** to measure shorter lengths.

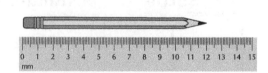

The length of this pencil is 12 centimetres.

We can also use **centimetres** to measure more accurately.

The length of this car is exactly 406 centimetres or 4 m and 6 cm.

1 m

1 m   1 m   1 m   1 m   1 m

Let's practise

1) What is the length of each of the bars below?

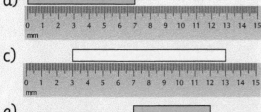

a)

b)

c)

d)

e)

f)

2) Estimate the length of each of the bars below then measure the actual lengths in centimetres and millimetres using a ruler.

Present your measurements in a table.

a)

b)

c)

d)

e)

f)

3) Isla measures Finlay's height by:

- getting him to stand up straight against a wall
- drawing a small pencil mark on the wall in line with the top of his head
- using a tape measure to measure from the floor to the pencil mark.

Choose three people in your class and measure their height in the same way. Record your results in your jotter.

CHALLENGE!

a) Use chalk or a length of string. Draw lines (or cut pieces of string) that you estimate to be:

- 10 cm
- 25 cm
- 1 m
- 3 m

b) Use a ruler, metre stick or tape measure to measure the lengths accurately and check how close you were.

c) Choose four or five objects in the environment around you and estimate their length. Measure accurately and check how close you were.

9 Measurement

9.3 Estimating and measuring mass

We are learning to estimate and measure the mass of an object.

Before we start

List three items in your class that:
a) weigh more than your school bag
b) weigh less than your school bag.

Mass is the measurement of how heavy an object is.

Let's learn

We can measure mass using scales, for example:

balance scales   kitchen scales   hanging scales   bathroom scales

We use **kilograms** to measure heavier objects.   We use **grams** to measure lighter objects.

We can also use **grams** to measure heavier objects more accurately.

The mass of this bag of shopping is about three kilograms.

The mass of this teddy bear is 150 grams.

The mass of the shopping bag is exactly 2900 grams.

Let's practise

1) What is the mass of each object below?

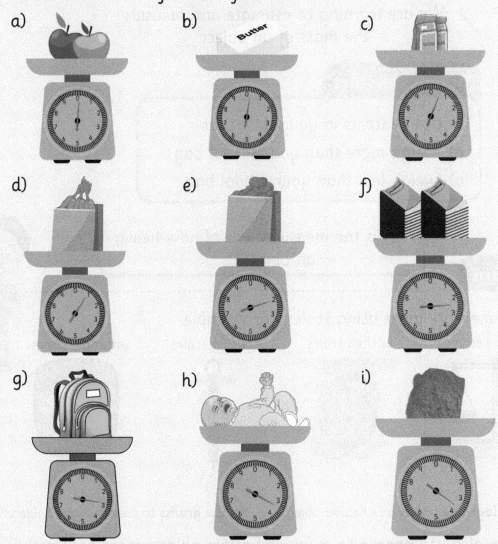

a)

b)

c)

d)

e)

f)

g)

h)

i)

2) Choose six of the objects listed below. Estimate the mass of each of the objects then measure the actual mass using a suitable set of scales. Copy and complete the table.

Item	Estimated Mass	Actual Mass
textbook		
pencil case		

backpack		
pupil		
drawer		
bottle of paint		
box of paperclips		
two die		
stapler		
pencil		
rubber		

CHALLENGE!

You will need:

- suitable scales (hanging scales may work best)
- a selection of items to be weighed
- a bag

a) Nuria has been challenged to fill a bag with items that total the following weights.

- 250 grams
- 500 grams
- one kilogram
- three kilograms

Help Nuria complete her challenge.

b) Choose some objects that you estimate will total each of the weights above. Try adding or removing objects to see how close you can get.

9.4 Converting units of length

We are learning to convert between metric units of length.

Before we start

Use a ruler to measure the length of each of the bars below. Give your answers in both centimetres and millimetres.

We can use **metres**, **centimetres** and **millimetres** to measure length.

Let's learn

'Centi' at the beginning of a word means 'one hundredth'. One metre equals one hundred centimetres.

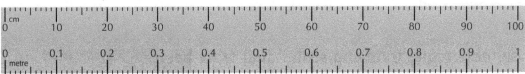

We can measure length in metres.

We can be more accurate by stating the length in centimetres or using decimals:

This table is about two metres long.

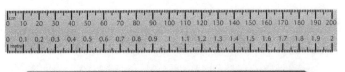

It is exactly 196 cm long or 1·96 m.

One centimetre equals 10 millimetres.

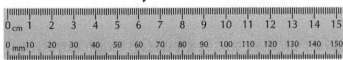

We can be more accurate by stating the length in millimetres or using decimals.

We can measure length in centimetres.

This golf tee is about three cm long.

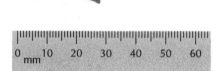

It is exactly 28 mm long or 2·8 cm.

Let's practise

1) The following measurements have been given in metres. Convert each of the measurements into centimetres.

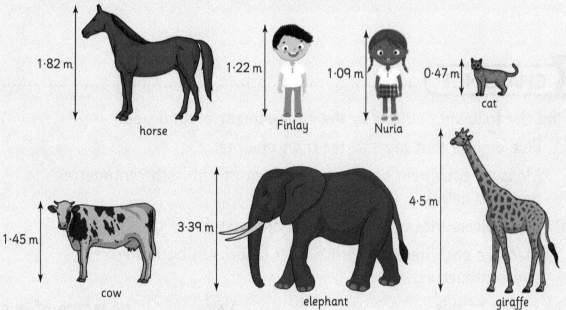

1·82 m — horse

1·22 m — Finlay

1·09 m — Nuria

0·47 m — cat

1·45 m — cow

3·39 m — elephant

4·5 m — giraffe

2) The following measurements have been given in centimetres. Convert each of the measurements into millimetres.

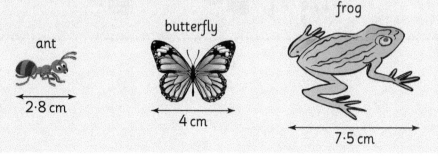

ant — 2·8 cm

butterfly — 4 cm

frog — 7·5 cm

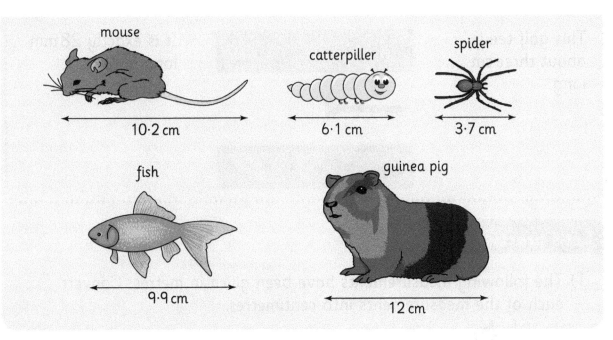

mouse
10·2 cm

catterpiller
6·1 cm

spider
3·7 cm

fish
9·9 cm

guinea pig
12 cm

⭐ **CHALLENGE!**

Find the following objects in the environment around you:

a) Five objects that are shorter than one metre.

Measure each item and write their lengths in both centimetres and millimetres.

b) Five objects that are longer than one metre.

Measure each item and write their lengths in both metres and centimetres.

The height of this chair is **47·2 cm** or **472 mm**.

47·2 cm

1·68 m

The length of this table is **1·68 m** or **168 cm**.

9

134   **Measurement**

9.5 Calculating the perimeter of simple shapes

We are learning to calculate the perimeter of regular shapes.

Before we start

Use a ruler to measure each of the lines below. Give your answers in both centimetres and millimetres.

The total distance around all of the sides of a 2D shape is called the **perimeter**.

Let's learn

The perimeter of a shape can be found by measuring and adding the total length of all the sides.

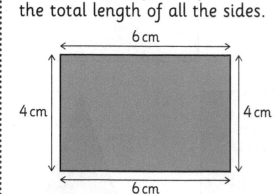

We can find the **perimeter** of this shape by adding up the lengths of all the sides:

6 + 4 + 6 + 4 = **20**

perimeter = 20 cm

Amman has been asked to measure the perimeter of his classroom. He's measured two sides so far.

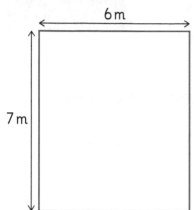

Amman says: 'I still need to measure the other two sides if I want to work out the perimeter.'

Isla says: 'The classroom is rectangular so opposite sides are the same length.'

The perimeter can be worked out without having to measure all the sides.

7 + 6 + 7 + 6 = 26 so the **perimeter = 26 m**

Let's practise

1) Calculate the perimeter of each of the shapes below.

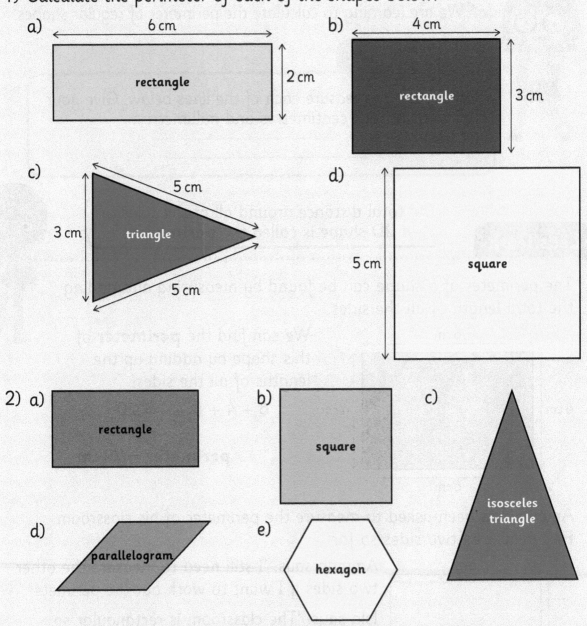

a) 6 cm   rectangle   2 cm

b) 4 cm   rectangle   3 cm

c) 5 cm   3 cm   triangle   5 cm

d) 5 cm   square

2) a) rectangle

b) square

c) isosceles triangle

d) parallelogram

e) hexagon

   i Use a ruler to take measurements and calculate the perimeter of each shape.

  ii What is the fewest number of measurements you will need to take for each shape?

iii Copy and complete the table.

Shape	No. of measurements	Perimeter
rectangle	2	4 + 2 + 4 + 2 = 12 cm
square		
isosceles triangle		
parallelogram		
hexagon		

⭐ **CHALLENGE!**

Choose four objects in the environment around you, for example, table, classroom, poster and jotter. Estimate the perimeter of each and then choose a suitable measuring device and calculate the perimeter.

Copy and complete the table below:

Object	Estimated perimeter	Actual perimeter

9.6 Finding the area of regular shapes in square cm/m

We are learning to calculate the area of regular shapes.

Before we start

Finlay lines up his toy soldiers in four rows of six in each row. How many soldiers are there altogether?

Amman has 20 toy soldiers. How many different ways can he line them up so that they are in equal rows?

Understanding how to multiply using an array will help us calculate the area of squares and rectangles.

Let's learn

The **length** of this line is one centimetre. $\overline{}$ 1 cm

This is the distance from one end of the line to the other: length measures one dimension.

Area is the space covered by a 2D shape. We measure area using **squares**.

This is a **square** centimetre.   1 cm ⬚
We can write **1 cm²**
                                    1 cm

It would take six squares to cover the rectangle. We can say the area is six square centimetres.

Area = 6 cm²

This rectangle is an array with two rows and three squares in each row so we can work out the area as:

2 cm × 3 cm = 6 cm²

Let's practise

1) Calculate the area of the shapes below.

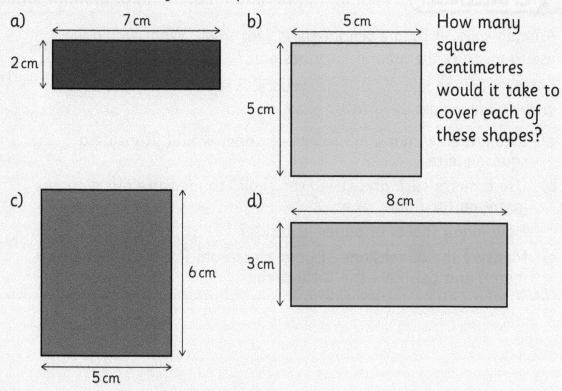

a)

7 cm

2 cm

b)

5 cm

5 cm

How many square centimetres would it take to cover each of these shapes?

c)

6 cm

5 cm

d)

8 cm

3 cm

2) Use a ruler to measure the shapes below and calculate their surface area.

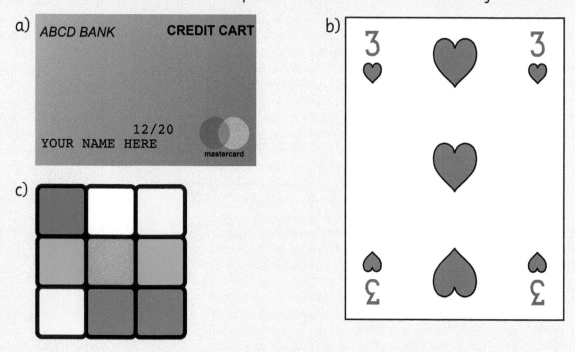

a)

ABCD BANK   CREDIT CART

12/20
YOUR NAME HERE

mastercard

b)

3
♥

♥

♥

3
♥

♥

♥

♥

c)

CHALLENGE!

Amman says: 'A square centimetre is very small. What would we use to measure the area of our classroom?'

1 m

1 m

Isla says: 'We should use a larger square. A square metre is a square where all four sides are a metre long.'

a) Estimate how many A4 sheets of paper would fit inside a square metre.

b) Use a metre stick and string (or chalk) to measure out a square metre on the floor.

    How many sheets of paper fit inside?

c) Measure the dimensions of your classroom floor (to the nearest metre) and calculate the total area.

9.7 Finding the volume of cubes and cuboids by counting cubes

We are learning to measure the capacity of a vessel.

Before we start

What is the volume of these 3D objects?

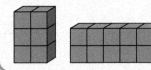

Capacity is the amount of liquid a container can hold. It is the volume of space taken up by a liquid.

Let's learn

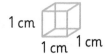

1 cm · 1 cm · 1 cm

A **millilitre** is the amount of liquid it would take to fill a cubic centimetre.

We use **millilitres** to measure smaller capacities.

A tablespoon holds 10 millilitres of water.

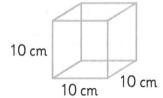

10 cm · 10 cm · 10 cm

A **litre** is the amount of liquid it would take to fill a cube with sides that measure 10 centimetres.

We use **litres** to measure larger capacities.

five litres

This bottle holds five litres of water.

We can also use **millilitres** to measure larger capacities more accurately.

4900 millilitres

The bottle contains 4900 millilitres of water.

Let's practise

1) Nuria is unsure how to read the scale on this measuring jug.

Amman says: 'There are five steps between the 0 and 500. 500 divided by 5 is 100. So each step must be worth 100.'

Isla says: 'The water is level with the fourth step. That means this must be 400 millilitres.'

How much liquid is there in each of the containers below?

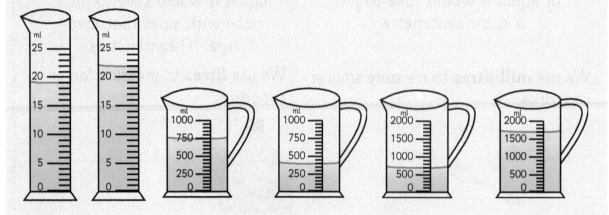

2) Select six different containers (plastic cups, bottles, jugs, etc.) and follow the instructions below.

a) Fill each container with a different amount of water.

b) Estimate the amount of water in each container.

c) Pour the water from each container into a measuring jug.

d) Read the measurement using the scale.

This measures about 600 ml.

e) Copy and complete the table:

Container	Estimate	Actual
1		
2		
3		
4		
5		
6		

CHALLENGE!

You will need:
- a two litre plastic bottle or jug
- a measuring jug
- water

How can you pour the following amounts into a water bottle without using a measuring jug?
- 50 millilitres
- 200 millilitres
- One litre
- 1500 millilitres

Once you have poured your guess, you can check by pouring it into the measuring jug.

Challenge a partner to pour out different capacities of water.

9.8 Estimating and measuring capacity

We are learning to find the volume of cubes and cuboids.

Before we start

Calculate the areas of these two rectangles.

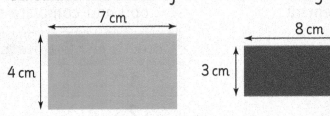

7 cm

4 cm

8 cm

3 cm

Volume is the amount of space taken up by an object.

Let's learn

The **area** of this square is one square centimetre (1 cm²).

1 cm

1 cm

This is the space covered by a two-dimensional shape.

Volume is the space taken up by a 3D object.
We measure volume using **cubes**.

This is a **cubic** centimetre.
We can write **1 cm³**.

1 cm

1 cm   1 cm

It would take six cubes to make this cuboid. We can say the volume is six cubic centimetres.

Volume = 6 cm³

A cubic centimetre is one cm long, one cm high and one cm deep.

Let's practise

1) Use cubic centimetres to make each of the cubes and cuboids below. What is the volume of each shape?

a)   b)    c) 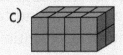   d)

2) Write down how to find the volume of the objects in question 1 **without** counting the cubes.

3) Find the volume of these objects **without** counting the cubes.
   Remember that some of the cubes are hidden.
   Check your answer by building the shapes afterwards. Were you correct?

a) 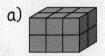   b)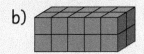

c) 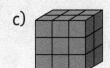   d) 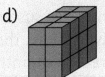   e)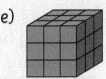

⭐ **CHALLENGE!**

The children each have some cubic centimetres. They have been asked to make as many cubes or cuboids as they can using all their cubes.

Amman	Isla	Finlay	Nuria

a) Predict who you think will be able to make the most shapes.

b) Investigate how many different cubes or cuboids each of the children can make using all of their cubic centimetres.

(You may wish to use blocks or multilink cubes to help you.)

10.1 Mathematical inventions and different number systems

We are learning to discuss important mathematical inventions and number systems.

Before we start

Discuss with a partner what you think a mathematician is.

Can you name any famous mathematicians?

What were the following mathematicians famous for?

Blaise Pascal

Leonardo Fibonacci

Numbers and number systems have been used by civilisations throughout history to record quantities.

Let's learn

A **number system** is a way of recording and expressing numbers using digits, letters or symbols in a consistent way.

The Romans used a system of numerical notation based on a combination of the following letters:

I V X L C D M

The most commonly used number system is the **Hindu-Arabic system,** which is a set of 10 symbols – **1, 2, 3, 4, 5, 6, 7, 8, 9** and **0**. This is the **decimal number system** that we use today. This was originally created by mathematicians in India in the sixth or seventh century.

Roman Numerals

1 = I	10 = X	100 = C	1000 = M
2 = II	20 = XX	200 = CC	2000 = MM
3 = III	30 = XXX	300 = CCC	3000 = MMM
4 = IV	40 = XL	400 = CD	
5 = V	50 = L	500 = D	
6 = VI	60 = LX	600 = DC	
7 = VII	70 = LXX	700 = DCC	
8 = VIII	80 = LXXX	800 = DCCC	
9 = IX	90 = XC	900 = CM	

Let's practise

1) Using the information from the table, write down the following numbers using Roman numerals:
   - a) 22
   - b) 31
   - c) 29
   - d) 45
   - e) 93
   - f) 126
   - g) 555
   - h) 2018

2) Using the information from the table, write down the following facts about yourself using Roman numerals, and ask a friend to check these:
   - a) Age
   - b) Date of birth
   - c) Shoe size
   - d) Number of people in your family

3) Copy and complete the following calculations using Roman numerals:
   - a) XXIX + _____ = XXXVIII
   - b) X + XI = _____
   - c) DCC – CCCL = _____
   - d) _____ + LXXV = CXXIII

CHALLENGE!

Let's investigate ...

John Napier is a famous Scottish mathematician.

Work with a partner to research what he is famous for and what impact he has had on the world.

Create a fact file with the information you find to include the following details:

a) His name and date of birth (you could include his nickname!)

b) Where in Scotland he was from

c) What he was best known for

d) Examples of his work (include pictures and diagrams).

11.1 Exploring and extending number sequences

We are learning to identify and extend a pattern or sequence.

Before we start

Continue the following pattern and explain the relationship you have identified.

★ ★ ● ★ ★ ● ★ ★   ?

Then create your own pattern and ask a friend to continue it.

Patterns can be created using relationships between drawings, letters and numbers.

Let's learn

The next shape in the sequence would be a ▲ as the pattern is two circles, then one triangle, then one circle, then one triangle repeated.

Drawing pattern – this is a pattern or sequence that contains a series of images, pictures, symbols rather than letters.

D, H, L, P, __ T This sequence contains every fourth letter in the alphabet.

Letter pattern – this is a pattern or sequence that contains a series of letters.

3, 8, 13, 18, 23, __ 28 The rule is 'add five'.

Number pattern – this is a series of numbers which follows a specific sequence or pattern that is called a **rule**.

Let's practise

1) Copy and complete the following patterns and sequences:

a)

b)

c) A, D, G, J, ☐

2) Copy and complete the following number patterns and state the rule:

a) 2, 3, 5, 8, 13, 21, _____, _____ The rule is _____

b) 1, 9, 17, 25, 33, 41, _____, _____ The rule is _____

c) 99, 92, 85, 78, 71, 64, _____, _____ The rule is _____

CHALLENGE!

1) Using these shapes create a repeating pattern that has twice as many squares as triangles.

2) Look at this number pattern:

4, 15, 26, 37, 48

This number pattern increases by multiples of 11 each time, so the rule is 'add 11'.

Create your own number pattern and copy and complete the sentence:

This number pattern _____

12 Expressions and equations

12.1 Solving simple equations using known number facts

We are learning to solve expressions and equations using +, −, ×, ÷.

Before we start

True or false?

a) $5 \times 12 = 6 \times 10$

b) $2 \times 25 > 3 \times 20$

c) $275 < 1000 - 850$

d) $10 \times 45 \neq 525 - 75$

Equations are number sentences which contain expressions that balance.

Let's learn

An **expression** is a combination of numbers, symbols and operators that are grouped together to show the value of something.

7 × 7   This expression shows a calculation for 49.

An **equation** is a statement that says that two things are **equal** and it will have an equals sign '=' in the middle.

7 × 7 = 49   This equation tells us that the value of completing the operation to the left of the equals sign is the **same** as the value of the number on the right of the equals sign.

Look at the following equation:

8 ✳ 4 = 32   What does the ✳ stand for?

What operation do you need to do to make 8 and 4 equal 32?

8 × 4 = 32   The answer is **multiply**.

Let's practise

1) Copy and complete the following equations and find what the ✳ stands for,

   a) 7 × ✳ = 35   b) 87 ✳ 22 = 65   c) 45 + 32 = ✳

2) Read the word problems and solve the number sentences:

   a) Isla helped her father to make some chairs.
      At the end of the week he had made 21 chairs.
      How many did he make each day?

      ⬜ × 7 = 21

   b) Nuria has five dresses.
      Her aunt makes her some more dresses and now she has eleven.
      How many dresses did her aunt make her?

      5 + ⬜ = 11

   c) Amman has three chickens.
      If each chicken lays one egg a day, how many days will it be until he has nine eggs?

      3 × ⬜ = 9

3) Read the word problems and solve the number sentences.

   a) A beetle has six legs. How many beetles have 30 legs?

      6 × ⬜ = 30

   b) A box has 42 chocolates arranged in rows of seven. How many rows are there?

      42 ÷ ⬜ = 7

   c) I think of a number, then I divide it by five. The answer is seven. What was my number?

      ⬜ ÷ 5 = 7

CHALLENGE!

For each number sentence, find the unknown.

a) 6 × ⬜ = 98 − 68   b) 545 − 425 = ⬜ × 12
c) 675 ÷ ⬜ = 5 × 45   d) 650 ÷ 10 = ⬜ + 43

13.1 Drawing 2D shapes

We are learning to draw squares and rectangles accurately.

Before we start

Isla says there are two right angles in this diagram.

Finlay says there are two acute angles and only one right angle.

Who is right? Use a protractor to justify your answer.

It is important to be accurate when drawing shapes. Always use a ruler for straight sides, and a protractor to measure angles.

Let's learn

How to draw a square with sides of length 6 cm.

1) Take a ruler and draw the base. It must be exactly 6 cm long.

2) Place your protractor on the right hand end of the base line. Find 90 degrees and make a mark.

3) Line up your ruler from the end of the base line to the mark. Measure 6 cm accurately, and draw your second side.

4) Put your protractor on the other end of the base line. Mark the 90 degree point.

5) Line up your ruler from the end of the base line to the mark and measure 6 cm exactly. Draw the third side.

6) Line your ruler up between the two sides you have just drawn. Check that there is exactly 6 cm between them. Draw the fourth side.

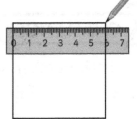

You can draw a rectangle the same way. Make sure the opposite sides are the same length.

Let's practise

1) Draw a square with sides of length 5 cm.

2) Draw a 2 cm × 8 cm rectangle.

3) Isla has created this design (not to scale). She started with a 4 cm square, then drew a 6 cm square, a 8 cm square and a 10 cm square.

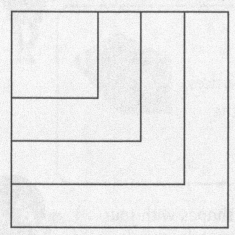

Copy Isla's design, then add two more squares.

4) This design has been created using 2 × 3 cm rectangles. Copy it, then add more rectangles to it to make your own design.

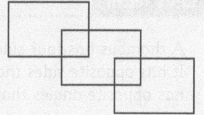

CHALLENGE!

Try drawing a rhombus with sides of length 5 cm.

13.2 Naming and sorting 2D shapes

> We are learning to identify, describe and sort 2D shapes by their properties.

Before we start

Nuria says this is a pentagon.

Amman says that this is **not** a pentagon, because pentagons have five sides and six vertices.

Finlay says it is **not** a pentagon because the sides are different lengths.

Who is right? Justify your answer.

> There are many types of shapes with four sides. These are all called quadrilaterals.

Let's learn

A rhombus has four sides of equal length. It has opposite sides that are parallel and it has opposite angles that are equal.

A trapezium has four sides, two of which are parallel. The sides may all be different lengths.

A parallelogram is any quadrilateral in which **both** pairs of opposite sides are parallel. Squares, rectangles and rhombuses are all special cases of parallelogram.

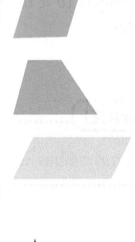

A kite has two pairs of equal sides, that are next to each other. It has one pair of equal angles opposite one another. A rhombus is a special case of kite.

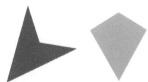

Let's practise

1) Nuria is sorting these shapes using the flow chart. Can you help her? Copy and complete the flow chart.

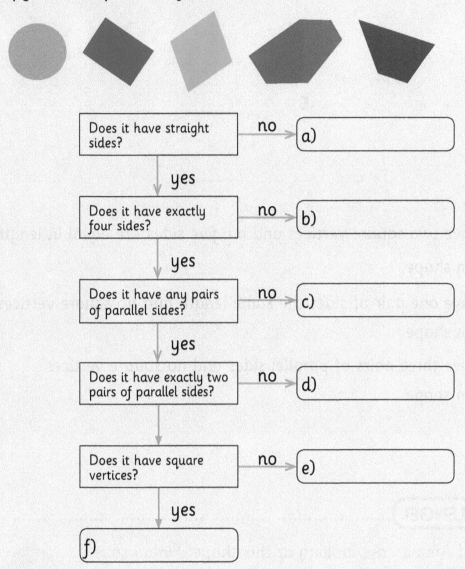

| Does it have straight sides? | no → | a) |

yes ↓

| Does it have exactly four sides? | no → | b) |

yes ↓

| Does it have any pairs of parallel sides? | no → | c) |

yes ↓

| Does it have exactly two pairs of parallel sides? | no → | d) |

↓

| Does it have square vertices? | no → | e) |

yes ↓

f)

2) Solve these puzzles about the shapes A to E.

A   B   C

D   E

a) I have two square vertices and my five sides are equal in length.

I am shape ⬚

b) I have one pair of sides the same length and no square vertices.

I am shape ⬚

c) I have three pairs of parallel sides and no square vertices.

I am shape ⬚

⭐ **CHALLENGE!** ....

Finlay and Amman are looking at this shape. Finlay says it's a kite. Amman thinks it's a trapezium. Is either of them correct? How do you know? Talk to a partner.

13 2D shapes and 3D objects

13.3 Drawing 3D objects

We are learning to draw cubes and cuboids.

Before we start

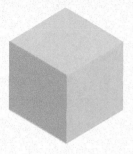 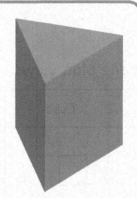

Which of these objects is the odd one out? Justify your answer.

When we draw cubes and cuboids we must make them look 3D even though they are only 2D on the page.

Let's learn

You will need 1 cm isometric (dotty) paper.

Using a ruler, draw a rhombus for a cube, or a parallelogram for a cuboid, by joining the dots.

Add three vertical lines at the vertices as shown.

Join the vertical lines at the bottom.

Cube

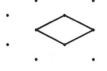

Cuboid

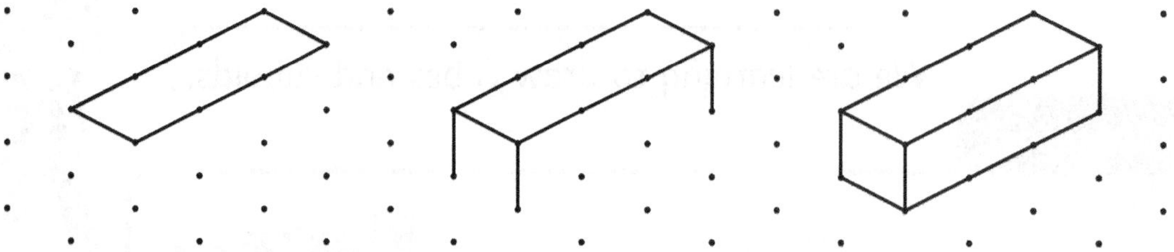

Here are the plan views of the cube and cuboid.

	Cube
Top view	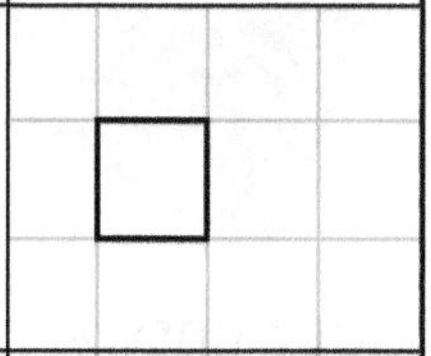
Side view	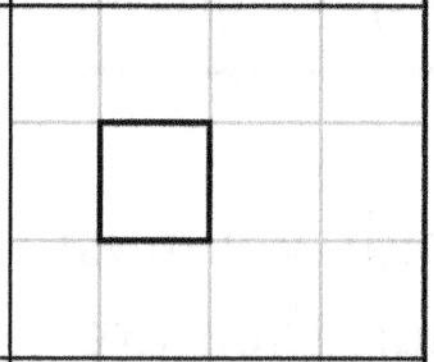
End view	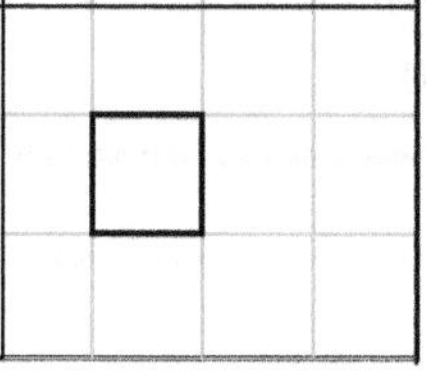

	Cuboid
Top view	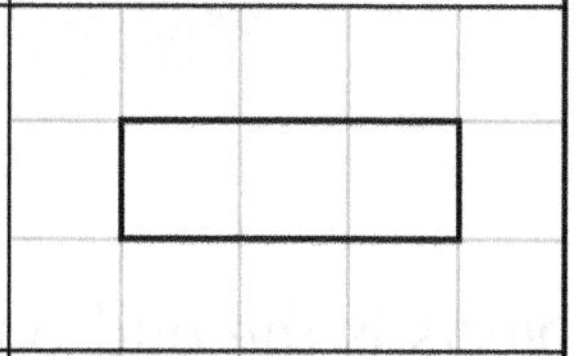
Side view	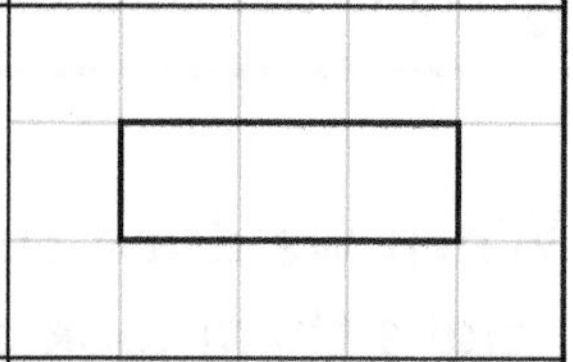
End view	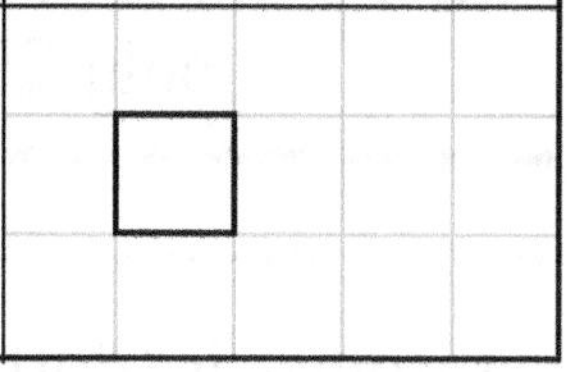

Let's practise

1) On isometric paper, draw:
   a) a cube with edges of length 2 cm
   b) a 4 cm × 4 cm × 2 cm cuboid
   c) a 6 cm × 5 cm × 4 cm cuboid.

2) On squared paper, draw a table like the one above, and draw the top, side and end views for each of your drawings in Question 1.

3) Draw the top view, two side views and two end views for each of objects A and B below:

A

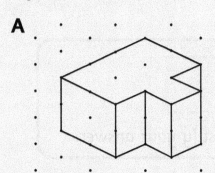

B

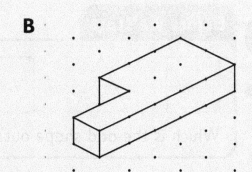

⭐ **CHALLENGE!**

Make some 3D objects using six interlocking cubes. Draw them on isometric paper.

One has been done for you.

Challenge a friend to make the objects you have drawn.

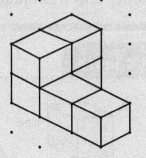

13.4 Describing and sorting prisms and pyramids

We are learning to identify, describe, sort and compare 3D objects.

Before we start

Which is the odd shape out? Justify your answer.

A prism has the same cross-section across its length. It has straight, flat sides and two identical end faces. The shape of the end face gives the prism its name.

A pyramid has a flat base, which can be any straight-sided shape, and triangular faces that meet at a vertex at the top.

Let's learn

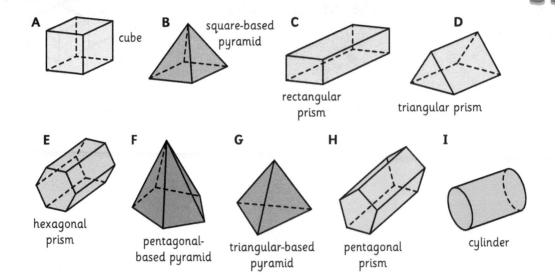

A   cube

B   square-based pyramid

C   rectangular prism

D   triangular prism

E   hexagonal prism

F   pentagonal-based pyramid

G   triangular-based pyramid

H   pentagonal prism

I   cylinder

The shape of the base gives the pyramid its name.

Let's practise

1) Complete the table. For the 3D objects A to I on the previous page, mark the prisms with a tick (✓) and the ones that are not prisms with a cross (✗).

3D object	A	B	C	D	E	F	G	H	I
prism	✓								
not a prism									

2) For objects A to H on the previous page, complete the table by counting:

a) the edges.   b) the vertices.

3D object	A	B	C	D	E	F	G	H
number of edges	12							
number of vertices	8							

3) Copy and complete the diagram by writing the letter of the 3D object from A to H in the correct region.

At least one triangular face

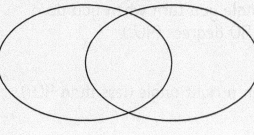

More than five vertices

CHALLENGE!

a) Is there a relationship between the number of vertices and the number of edges of a prism? Investigate.
   Hint: Look at your answers to question 2.

b) Write down what you have found out.

14.1 Identifying angles

We are learning to identify angles as acute, right, obtuse or straight.

Before we start

Mr Smith is stuck in the maze.
What does he need to do next?
a) Turn 90° clockwise
b) Turn 180°
c) Walk forward
d) Make a quarter turn anticlockwise

Angles can be many different sizes. We can sort angles into categories by comparing them to a right angle.

Let's learn

A right angle is the angle at the corner of a rectangular piece of paper. It is the angle you turn when you do a quarter turn. It measures 90 degrees (90°).

right

An acute angle is less than a right angle (less than 90°).

acute

An obtuse angle is greater than a right angle (more than 90°).

obtuse

A straight angle is where the arms of the angle are in a straight line. It measures exactly 180°.

straight

Let's practise

1) Are these angles **right**, **obtuse** or **acute**?

a)

b)

c)

d)

e)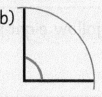

f)

2) Complete the table. Identify the angles as acute, right or obtuse.

Letter	Angle
A	
B	
C	
D	
E	
F	
G	
H	
I	
J	

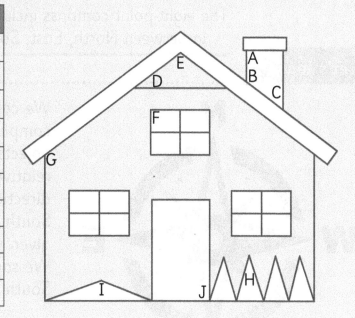

⭐ **CHALLENGE!**

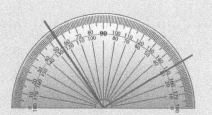

Here is an angle on a protractor.
Finlay thinks the angle is obtuse.
Isla thinks it is acute.
Amman thinks it is a right angle.
Who is right? Explain your answer.

14.2 Using an eight-point compass

We are learning to follow eight-point compass directions.

Before we start

Isla is facing West. Amman gives her these instructions:
Turn 90° clockwise, then turn 180°, then turn 90° anticlockwise.

What direction is Isla facing now? Explain your thinking.

The eight-point compass includes the directions in between North, East, South and West.

Let's learn

We can use the eight-point compass directions to locate objects or places on a map, relative to other objects. The directions in between North, South, East and West are always given with North or South first. We say North East, North West, South East or South West.

Let's practise

1) Amman made this plan of his home town.

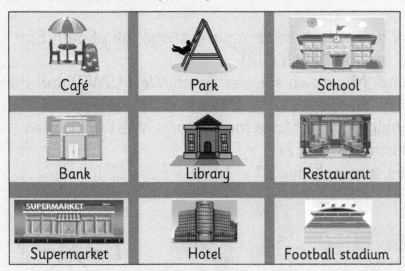

a) You are at the library. What is:

   i) to the North West?   ii) to the South East?

   iii) to the South West?   iv) to the North East?

b) Which building is:

   i) North East of the supermarket?   ii) South West of the restaurant

   iii) South East of the bank?   iv) North West of the hotel?

c) What two buildings are South West of the school?

2) Copy this grid into your jotter. Follow these directions and draw in colour the pathway that you take. The first section of the pathway has been completed for you.

   S4 – W2 – SE2 – E2 – NE2 –
   W3 – N1 – E2 – NW3

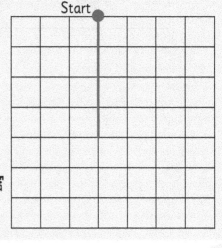

3) Follow the directions and write down all the places the squirrels have buried their nuts.

a) Start at the fox hole. Move three squares diagonally South East (SE3) then two squares North (N2).

b) Start at the pond. Move two squares South West (SW2) and then one square North West (NW1).

c) Start at the rabbit warren. Move four squares West (W4) then two squares North East (NE2).

d) Start at the tree stump. Move E1 – NW2 – E2 – SE1.

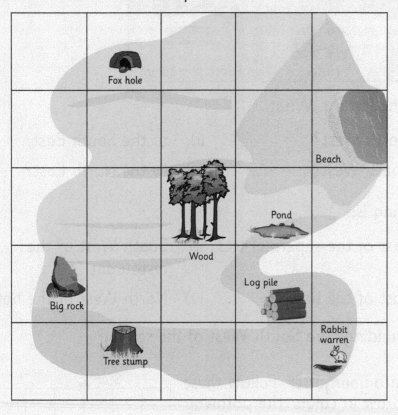

CHALLENGE!

Starting at A, there are many different eight-point compass directions you could give to get to B.

You must not cross where you have been before, or retrace your steps.

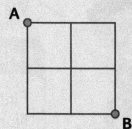

Not allowed:

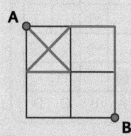

SE1-W1-NE1-E1-S2

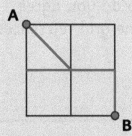

SE1-W1-E2-S1

This example is **allowed**:

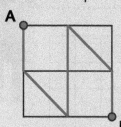

S1-SE1-N2-SE1-S1

1) What is the longest set of instructions you can write to get from A to B?

2) What if the grid was a rectangle three squares along and two squares up?

14.3 Plotting points using coordinates

We are learning to use coordinates to plot points on a graph.

Before we start

Finlay says the triangle is at C4, but Nuria disagrees. Who do you agree with and why? List the grid references of the other shapes.

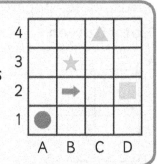

Coordinates describe exactly where a point is on a graph.

Let's learn

On a coordinate grid, the lines are labelled, not the boxes.

Coordinates are written like this: (3, 5).

To plot point (3, 5), you count from 0 to 3 horizontally, then from 0 to 5 vertically. Plot the point by marking an X.

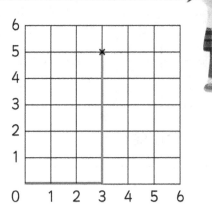

Let's practise

1) On squared paper, copy a grid like this.

On the grid, plot the following points:
a) (1, 2) (2, 4) (3, 6)
   What do you notice?
b) (1, 1) (5, 3) (5, 1)
   Join the dots with straight lines. What do you notice?

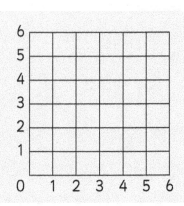

2) a) Copy another grid. Plot the points (1, 2) (1, 5) and (4, 5). Plot a fourth point to make a square. What are the coordinates of the fourth point?

   b) Copy another grid. Plot the points (1, 2) (3, 5) and (5, 4). Plot a fourth point to make a rectangle. Write down the coordinates of the fourth point.

3) Copy this grid.

   a) Plot these points, and join them up as you go along to make a letter of the alphabet:
   (4, 1) (4, 3) (3, 2) (2, 3) (2, 1) (1, 1) (1, 5) (2, 5) (3, 3) (4, 5) (5, 5)

   b) Write down the coordinates of the final point needed to complete the letter.

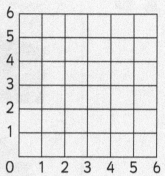

⭐ **CHALLENGE!** ....

Copy this grid.

Plot these 12 points: (0, 3) (8, 3) (4, 7) (8, 6) (6, 3) (5, 6) (1, 7) (3, 6) (3, 0) (1, 4) (4, 4) (5, 3)

These points form the corners of three squares. Work out where the three squares are. Draw the edges of the squares. Use a different colour for each square.

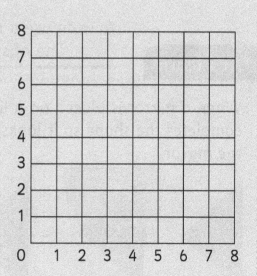

14.4 Lines of symmetry

We are learning to identify lines of symmetry in shapes and to complete shapes using symmetry.

Before we start

Which is the odd one out? Justify your answer.

A

B

C

D

You can use a mirror to help you identify a line of symmetry on a shape.

Let's learn

When a mirror is placed on a line of symmetry, the reflection completes the shape so it looks exactly the same as without the mirror.

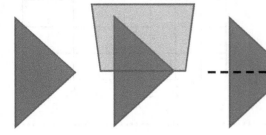

Line of symmetry

When you complete a picture to make it symmetrical, you can use a mirror to check.

Let's practise

1) In which of these shapes is the dotted line a line of symmetry?

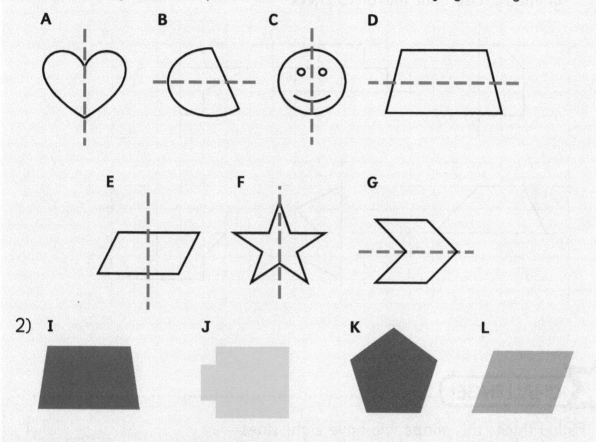

A   B   C   D

E   F   G

2) I   J   K   L

M   N   O   P

a) Use a mirror to test if there is a line of symmetry on each shape.
b) Copy and complete the table to sort the shapes.

Horizontal line of symmetry	Vertical line of symmetry	No line of symmetry

3) Copy and complete each shape by drawing the reflection in the line of symmetry. Use your mirror to check.

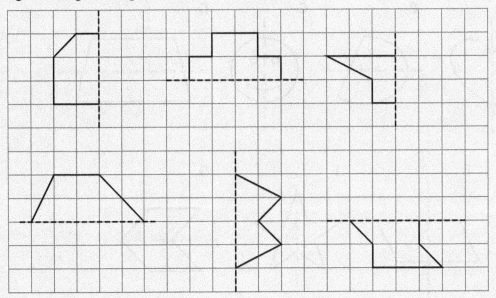

⭐ **CHALLENGE!** ....

Finlay thinks this shape will have eight sides once it is reflected in the line of symmetry.

Amman thinks it will have seven.

Who is right? Prove it.

14.5 Creating designs with lines of symmetry

We are learning to create designs with one line of symmetry.

Before we start

Work with a partner.

Take it in turns to make a design using shapes.

Ask your partner to use a set of the same shapes to make a symmetrical picture.

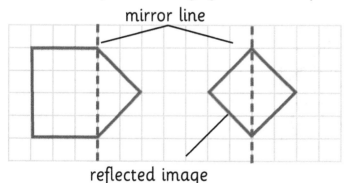

For a design to be **symmetrical**, shapes and colours have to be reflected in the line of symmetry.

Let's learn

A mirror can have two reflective faces.

Investigate the result when the mirror line crosses the shape in different places.

Draw the reflected image for each shape.

mirror line

reflected image

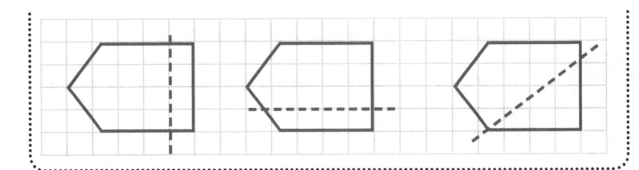

Let's practise

1) Use your mirror to see the reflected shape.
Copy and complete each pattern by drawing the reflection.

a)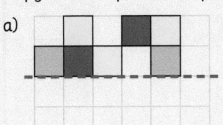

b)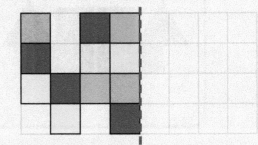

2) This quilt design is only half-made.
Reflect the pattern to complete the quilt design.

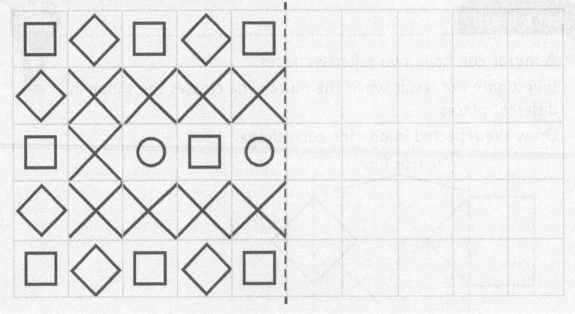

3) Draw your own design for a quilt.
   The quilt design must be symmetrical at the line of symmetry.
   Colour your quilt design, making sure that it is still symmetrical.

CHALLENGE!

Using exactly three colours, how many ways
can you colour these squares to make different
symmetrical designs?

14 Angles, symmetry and transformation

14.6 Measuring angles up to 180°

> We are learning to use a protractor to measure angles up to 180°.

Before we start

Here are some letters of the alphabet:

E F H I K L N T V W

Find the letters that have exactly:

One obtuse angle and two acute angles

One straight angle

Two right angles

Put them in order to make a three-letter word that completes the phrase P. E. ___ ___ ___

> We use a protractor to measure exactly how many degrees there are in an angle.

Let's learn

How to measure an angle:

Put the 'zero line' of the protractor on one arm of the angle (AB).

Make sure the vertex of the angle (B) is on the centre point of the protractor.

Measure the angle of the turn from the arm on the zero line (AB) to the other arm (BC).

Read the position of the other arm (BC) on the scale to find the angle.

Check: for example, if the angle is acute, did you measure less than 90°?

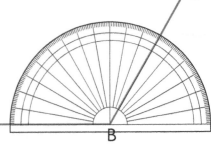

Let's practise

1) Use a protractor to measure each angle.

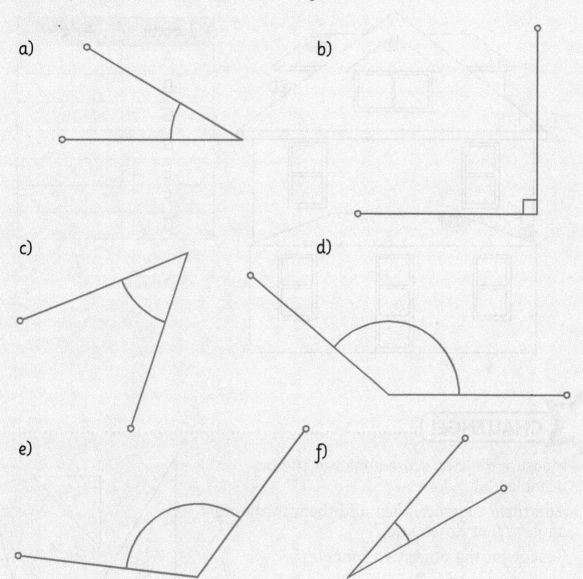

a)

b)

c)

d)

e)

f)

2) Measure the angles in the house.

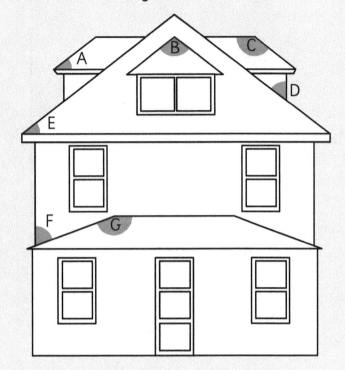

Letter	Angle
A	
B	
C	
D	
E	
F	
G	

⭐ **CHALLENGE!** ....

Measure the two acute angles in the big triangle.

Add them together, then add them to the right angle. What do you get?

Repeat for the other two triangles.

Think of a rule to describe how many degrees there are in a right-angled triangle. Describe your rule to a partner.

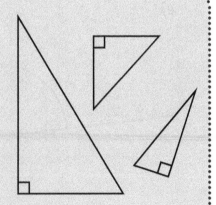

14 Angles, symmetry and transformation

14.7 Understanding scale

We are learning to understand scale.

Before we start

Amman's house is 2 km from school. He tells Nuria that he lives 2000 m away from school. Nuria thinks he lives 200 m from school. Who is right, and why?

We use scale to make sure that distances can be measured accurately.

Let's learn

The real-life distance between Bigtown and Smalltown is 10 kilometres. To draw this on a map it has been scaled down. On the map 10 **kilometres** has been represented as a 10 **centimetre line**.

This means that for every one centimetre on the map, the real-life distance is one kilometre.

We say the scale is 1 cm for every 1 km.

Smalltown

Bigtown

Let's practise

1) The scale on this map is 1 cm for every 1 km.

   How far is it in real life
   a) from Stirling to Blairlogie?
   b) from Blairlogie to Stirling, via Bridge of Allan?

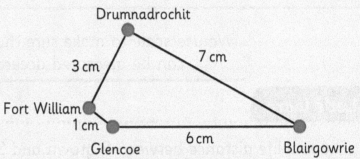

2) The scale on this map is 1 cm for every 25 km.

   How far is it in real life
   a) from Fort William to Glencoe?
   b) from Fort William to Drumnadrochit?
   c) from Blairgowrie to Fort William, via Drumnadrochit?

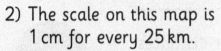

3) Amman is reading a map of Spain. The scale is 1 cm for every 50 km.

   On Amman's map,
   a) Bilbao is 8 cm from Madrid. What is the actual distance between Bilbao and Madrid?
   b) Malaga is 4 cm from Seville. What is the actual distance between Malaga and Seville?
   c) Valencia is 7 cm from Barcelona. What is the actual distance between Valencia and Barcelona?

CHALLENGE!

The scale on this map is 1 cm for every 40 km.

Use a ruler to measure the distances to the nearest centimetre, then answer the questions.

a) Which town is about 80 km east from Oban?

b) About how far is Ullapool from Inverness?

c) How far is it from Dumfries to Dundee, via Stirling?

d) Find two cities which are about 160 km from Ullapool.

15 Data handling and analysis

15.1 Reading and interpreting information

We are learning to read and make sense of data presented in graphs.

Before we start

The children are looking at this graph.

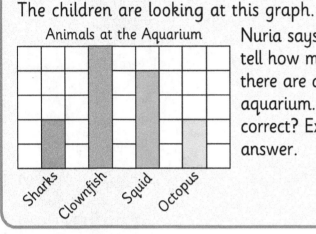

Animals at the Aquarium

Sharks · Clownfish · Squid · Octopus

Nuria says she can't tell how many animals there are at the aquarium. Is she correct? Explain your answer.

Graphs help us to understand data. They help us to compare the data and to draw conclusions about it.

Let's learn

We can use graphs to answer questions about real-life situations and the data gathered from them.

Bar graphs are useful to show relative sizes – we can see at a glance which sports are most liked, or least liked.

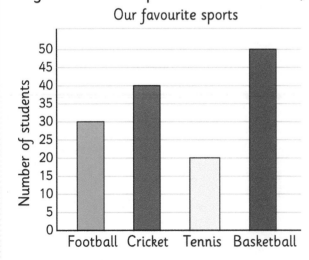

Our favourite sports

Number of students

Football · Cricket · Tennis · Basketball

Pictographs represent data in pictures. Each picture represents an amount. Here each television represents five televisions sold.

Day	Number of televisions sold
Day 1	▣ ▣ ▣
Day 2	▣ ▣
Day 3	▣ ▣ ▣ ▣ ▣ ▣
Day 4	▣ ▣ ▣
Day 5	▣

▣ – five televisions

Line graphs tell a story of changes over time.

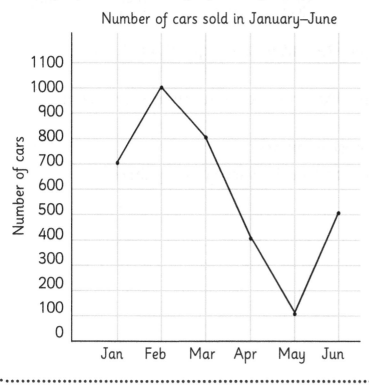

Number of cars sold in January–June

Let's practise

1) The bar graph shows the number of people who prefer different flavours of crisps. Look at the graph and answer the questions.

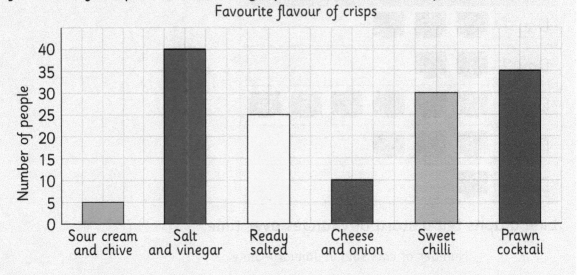

Favourite flavour of crisps

a) How many more people like salt and vinegar than cheese and onion?

b) How many more people like prawn cocktail than sour cream and chive?

c) Which flavour is the most popular?

d) How many people took part in the survey altogether?

2) The picture graph shows how many mobile phones a company sold over five days. Look at the graph and answer the questions.

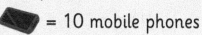 = 10 mobile phones

Day	Number of mobile phones sold
Day 1	🖥️ 🖥️
Day 2	🖥️
Day 3	🖥️ 🖥️ 🖥️ 🖥️
Day 4	🖥️ 🖥️ 🖥️ 🖥️ 🖥️
Day 5	🖥️ 🖥️ 🖥️ 🖥️

a) How many more phones did the company sell on Day 4 than Day 2?

b) How many more phones did the company sell on Day 5 than Day 1?

c) On which day did the company sell the least number of phones?

d) Over the five days, how many phones did the company sell altogether?

3) The line graph shows the number of laptops that a company sold over six months. Look at the graph and answer the questions.

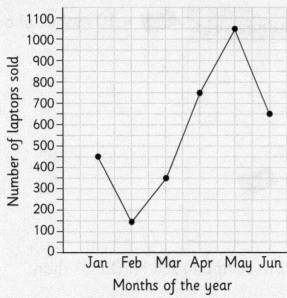

a) How many more laptops were sold in May than April?
b) How many more laptops were sold in June than February?
c) How many more laptops were sold in January than March?
d) Over the six months, how many laptops were sold altogether?

CHALLENGE!

Nuria has been growing a sunflower from seed. She has measured the sunflower's height every day for a week.

What is the best way for her to display her data?

Nuria wants to answer these two questions:

How much did the sunflower grow between Monday and Sunday?

Between which two days did the sunflower show the most growth?

Which of these questions would be easier for her to answer a) by looking at the graph and b) by looking at the table? Explain why.

Day	Height
Monday	13 cm
Tuesday	15 cm
Wednesday	19 cm
Thursday	22 cm
Friday	27 cm
Saturday	35 cm
Sunday	38 cm

15 Data handling and analysis

15.2 Organising and displaying data

We are learning to organise and display data in tables and graphs.

Before we start

Finlay has been watching the birds come to his bird feeder for one hour. He has made this table showing what he saw.

He wants to make a block graph. Isla says he should label both his axes 'BIRDS SEEN'. Is she correct? Explain your answer.

Blackbird	2
Sparrow	7
Blue tit	3
Goldfinch	4

When we collect data it is helpful to organise it, so we can read it, make sense of it, and display it in the best way.

Let's learn

Doctor's name	Number of patients seen on Monday (frequency)
Dr Finlayson	18
Dr Mortimer	20
Dr Martin	24
Dr Johnson	18

Frequency tables help us to organise data, so it is more useful. Once the data is organised into a frequency table, we read it more clearly. Frequency tables also make it easier to display the data in graph form.

Let's practise

1) A school wants to encourage healthy snacks. It conducts a survey into the kinds of snacks the children bring to school. The data is put into a table.

Complete the table by counting the tallies. One has been done for you.

Snack	Tally	Frequency
Crisps	ЖЖ ЖЖ ЖЖ ЖЖ ЖЖ ЖЖ ЖЖ I	36
Fruit	ЖЖ ЖЖ ЖЖ I	
Sweets	ЖЖ ЖЖ ЖЖ ЖЖ ЖЖ ЖЖ II	
Biscuits	ЖЖ ЖЖ II	
Vegetable sticks	ЖЖ III	

a) How many children bring fruit or vegetable sticks to school?
b) How many children bring sweets or biscuits to school?
c) How many more children bring sweets or biscuits than bring fruit?
d) Display the data in the table as a bar graph. Copy these axes. Label your axes and give your graph a suitable title.

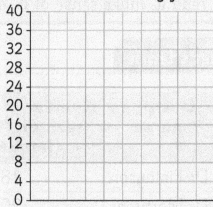

2) Isla collects information about the ages of children who attended her youth club on Thursday. She has written down all the ages on a piece of paper: Help Isla organise this information into the frequency table. Copy it into your jotter and complete.

8 10 7 6   8 7
10 7 8 6   8 6
8 8 10 10 7 7
8 8 8 7 10 8
10 6 6

Age in years	Tally	Number of children (frequency)
6	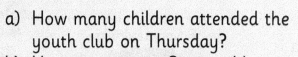	5
7		
8		
9		
10		

Display the data as a bar graph, using the scale shown. Label your axes and give your graph a suitable title.

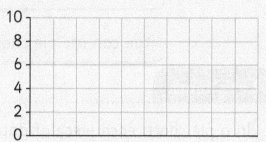

a) How many children attended the youth club on Thursday?
b) How many more 8-year-olds attended than 6-year-olds?
c) Which age group had the least number of children attending?

3) Ask each of the children in your class what their favourite playground games are.

Organise your data into a frequency table.

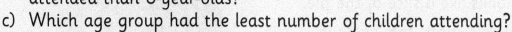

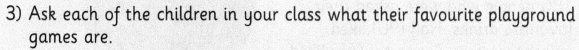

CHALLENGE!

Find out the favourite foods of children in your class or school by conducting your own survey.

Before you start, you might want to decide which categories you want to include: pizza, burgers, etc.

Create a frequency table of your results.

Display your data as a bar graph. Choose an appropriate scale for your axes and label them. Give your graph a suitable title.

15.3 Reading and interpreting pie charts

We are learning to read and make sense of data presented in pie charts.

Before we start

Isla's mum tells her to share her 32 strawberries fairly with her three friends. She gives Amman eight strawberries. 'Hey!' says Amman, 'that's not fair!' Is he right? How many strawberries should each friend receive?

You can use pie charts to display and compare data.

Let's learn

In a pie chart, each piece of data is represented as part of a whole, and looks like a slice of the pie.

In this pie chart, the whole is 180 people who were asked their favourite drinks. Half (90) liked milkshakes best. A quarter (45) liked water and a quarter liked juice.

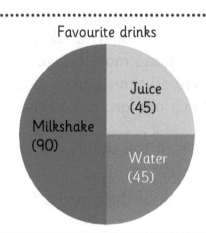

Favourite drinks

Let's practise

1) A shopkeeper gathers information about the flavours of ice-cream they sold in one day. The data is displayed in this pie chart.
   a) How many ice-creams were sold on that day?
   b) What was the most popular flavour?
   c) What was the least popular flavour?
   d) How many more people bought chocolate than pistachio?

Ice-cream flavours sold

2) This pie chart shows the favourite sports of
48 children at an after-school club.

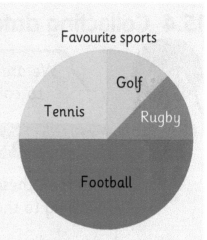

Favourite sports

a) How many children chose:
   i) football?   ii) rugby?
   iii) tennis?   iv) golf?
b) What was the most popular sport?
c) How many more children like football
   than golf?

3) A gardener has planted some vegetables.

Here is a frequency table showing the number of each type of
vegetable she planted.

Type of vegetable	Number planted (frequency)
Potato	28
Carrot	14
Onion	14

Draw a pie chart to display this data.
Remember to make sure your angles are correct.

CHALLENGE!

Write down two things that are the same about the two pie charts below.

Write down three things that
are different.

What do you think the pie charts
might tell us about how far
away the children live from
school A and school B?

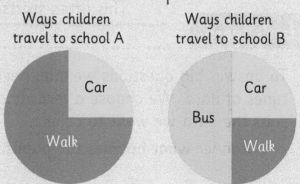

Ways children
travel to school A

Ways children
travel to school B

15.4 Collecting data

> We are learning to plan an inquiry, and to collect data in an organised way.

Before we start

Interpret these line graphs to say which best represents someone walking to the post box and back? Explain your answer.

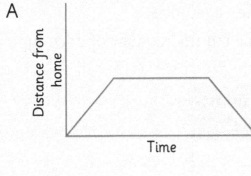

A — Distance from home / Time

B — Distance from home / Time

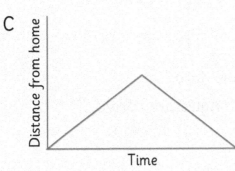

C — Distance from home / Time

> Inquiries usually start with a 'big question'. This often starts with 'I wonder ...'

Let's learn

To answer big questions we may need to collect different types of data. We choose a recording table or graph which suits the data we want to collect.

'I wonder what hobbies the pupils in my class do?'

For this type of question we collect **category data.** The choices of answers fit into categories such as sports, computer games or painting. Category data can be recorded in a list, a table or a bar graph.

Favourite flavour of ice-cream	Number of students
Vanilla	46
Strawberry	41
Chocolate	35
Pistachio	27
Toffee	31

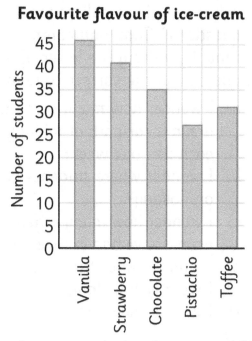

Favourite flavour of ice-cream

'I wonder how the height of a bean plant changes each day for a month?'

To answer this, we collect **time series data.** It is called time series data because we collect measurements regularly over a period of time. We can record time series data in a table or line graph.

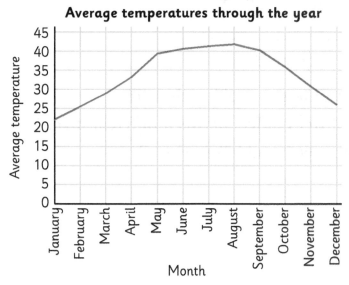

Average temperatures through the year

Day	Temperature (°C)
Jan	23
Feb	25
Mar	28
Apr	33
May	38
Jun	40
Jul	41
Aug	42
Sep	39
Oct	36
Nov	31
Dec	26

1) Some students have written these 'big questions'. What types of data will the students collect for these questions? Copy and complete the table by writing the letter of each question in the correct column.

Category data (words)	Category data (numbers)	Time series data

   a) I wonder how many cars there are in a car park at midday on each day one week?
   b) I wonder where students go on holiday?
   c) I wonder how many different jobs people have had in their lives?
   d) I wonder what the most popular websites are?
   e) I wonder how long it takes for an ice cube to melt?
   f) I wonder how many hours of sleep everyone got last night?

2) Match each inquiry to the correct type of graph to show the data. You can match more than one inquiry to the same graph.

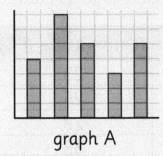

graph A

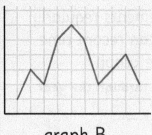

graph B

I wonder how many concert tickets are sold each day?

I wonder how the number of trees differs between different areas?

I wonder which famous places we would like to visit?

3) Use the templates to create a table for recording data for each inquiry.

a) Big question:
'I wonder how many cars people have owned in their lives?'
Categories:
0, 1–2, 3–4, 5–6, 7 or more

b) Big question:
'I wonder how my height will change if I measure it on 1st January each year?'
Age range:
10 to 14 years of age
Remember to include a unit of measurement in the table.

CHALLENGE!

Plan your own inquiry.

a) Think of a big question. Start with 'I wonder …'

b) Decide whether you will be collecting category data (numbers), category data (words) or time series data.

c) Make a table to collect your data. Give it suitable headings, including units of measurement if needed.

d) Carry out your survey or investigation to complete the table.

e) Make a bar graph or line graph to display your data. Remember to include a suitable title and to label your axes.

16 Ideas of chance and uncertainty

16.1 Predicting and explaining simple chance situations

We are learning to predict and explain the outcomes of chance situations.

Before we start

Look at the statements below and decide which is most suitable, and then discuss your answers with a partner.

impossible, **unlikely**, **even chance**, **likely** or **certain**.

a) There are 12 months in a year.

b) I will roll an even number on a dice.

c) Tomorrow I will be younger.

d) Someone in my class will have the same shoe size as me.

There are a number of different things that happen every day by chance.

Let's learn

When we talk about the **chance** or the **probability** of something happening we mean the **likelihood** of this event actually taking place.

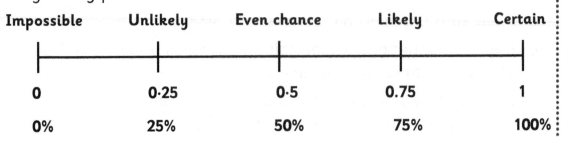

Impossible	Unlikely	Even chance	Likely	Certain
0	0·25	0·5	0.75	1
0%	25%	50%	75%	100%

There are some things that we know are **certain**:

there are seven days in a week.

There are some things that we know are **unlikely**:

it will never rain again.

Let's practise

1) Play 'Probability Stepping Stones, with a partner.

Place a counter on START and take turns to move one stone at a time.

Choose possible events only to get to the FINISH.

If you land on an impossible event you miss a turn.

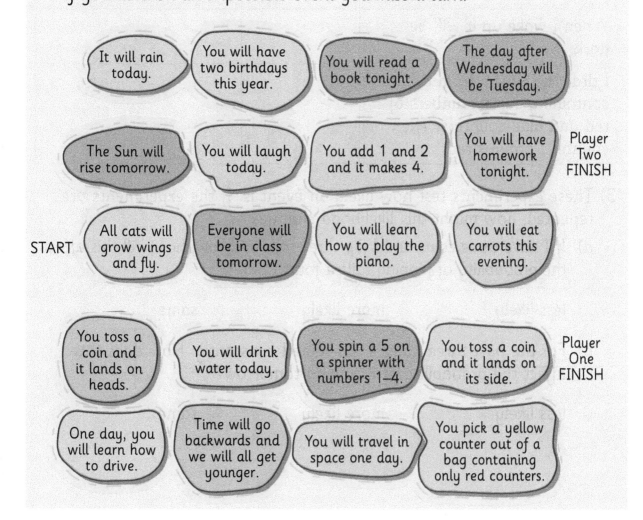

2) Using the probability scale, copy and complete the following table:

Impossible	Unlikely	Even chance	Likely	Certain
0	0·25	0·5	0·75	1
0%	25%	50%	75%	100%

Event	Probability	Explanation
I will go to school tomorrow.		
I will be older next week.		
I like ice-cream.		
When I wake up it will be dark.		
I draw a counter from a bag containing equal numbers of red and blue counters. The counter I draw is red.		

3) These experiments test how likely an event is. If the experiments are repeated, how might this likelihood change?

a) If I have three coins and I flip three tails in a row, how will this affect the probability of flipping tails a fourth time?

less likely ☐   more likely ☐   same ☐

b) For the past three days it has rained constantly. How does this affect the probability of it raining tomorrow?

less likely ☐   more likely ☐   same ☐

CHALLENGE!

1) A coin is flipped two times. List all the possible outcomes in the form H-T where H is heads and T is tails. How many possible outcomes are there?

2) A coin is flipped three times. List all the possible outcomes. How many possible outcomes are there?

3) A coin is flipped four times. List all the possible outcomes. How many possible outcomes are there?

4) Now, without listing them, how many possible outcomes are there for a coin flipped five times? Explain.

Answers

1 Estimation and rounding

1.1 Rounding whole numbers to the nearest 10 (p.2)

Before we start

Five thousand four hundred and ninety-two.
4 = 4 hundreds (400); 9 = 9 tens (90); 5 thousands;
1000 more = 6492

Questions

1) a) 30   b) 20   c) 30   d) 20
   e) 110   f) 120   g) 120   h) 120
   i) 540   j) 550   k) 540   l) 550

2) a) True   b) False   c) False
   d) False   e) True

3) a) 6
   b) 360, because 356 is closer to 360 than 350, so we round up.
   c) Any of 355, 357, 358, 359, 361, 362, 363 or 364

Challenge

Answers will vary.

1.2 Rounding whole numbers to the nearest 100 (p.4)

Before we start

5912

Questions

1) b) 900 and 1000 (rounded down)
   c) 600 and 700 (rounded down)
   d) 800 and 900 (rounded up)

2) b) False   c) True   d) True   e) False

3)

Number of visitors	May	June	July	August	September
Actual numbers	793	996	2097	1071	617
Rounded to nearest hundred	800	1000	2100	1100	600

Challenge

a) 1345, 1346, 1347, 1348, 1349

b) 1185

c) Answers will vary.

1.3 Rounding decimal fractions to the nearest whole number (p.6)

Before we start

0.6; one tenth more = 0.7; one tenth less = 0.5

Questions

1) a)

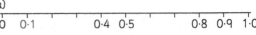

   b) 0.1 = 0; 0.4 = 0; 0.5 = 1; 0.8 = 1; 0.9 = 1

2) a) 5   b) 7   c) 5   d) 6   e) 5

3)

Month	Warmest (°C)	Round to nearest whole number	Coldest (°C)	Round to nearest whole number
January	6·6	7	0·5	1
February	7·1	7	0·6	1
March	9·1	9	1·9	2
April	11·3	11	3·3	3
May	14·4	14	5·8	6
June	17·3	17	8·7	9
July	19·3	19	10·6	11
August	19·1	19	10·5	11
September	16·4	16	8·5	9
October	13·0	13	5·8	6
November	9·2	9	2·5	3
December	7·2	7	1·1	1

Challenge

Answers will vary.

1.4 Estimating the answer using rounding (p.8)

Before we start

No, 1547 to the nearest 10 is 1550.

Questions

1) a) 30 + 60 = 90
   b) 80 + 60 = 140
   c) 30 + 90 = 120

2) a) Isla rounded the two numbers to 30 and 210, giving a total of 240, so 461 is clearly wrong. If you write the numbers underneath each other to add them, you should be able to see where 461 came from.

b) Finlay rounded 96 to 100 and doubled it to make 200.

Challenge
Rounding for the flowers, the number is roughly (330 + 420) × 2 = 1500.

And for the balloons (270 + 330) × 2 = 1200.

So the answers to both calculations are reasonable.

2 Number – order and place value

2.1 Reading and writing whole numbers (p.10)

Before we start
Finlay is incorrect because he has included an extra zero and has written seven thousand and five. He should have written 705. The zero in the tens place means there are no tens.

Questions
1) a) Three thousand, one hundred and thirty-eight
   b) Six thousand, seven hundred and twenty-four
   c) Seven thousand and two
   d) Five thousand, five hundred

2) a) 2078 The place holder is in the hundreds position.
   b) 5302 The place holder is in the tens position.
   c) 6160 The place holder is in the ones position.
   d) 9099 The place holder is in the hundreds position.

3) Answers will vary.

Challenge
Nuria is incorrect because she has forgotten to write the word 'and' before the word seventeen.

Isla is incorrect because she has written seventy instead of seventeen.

The girls should have written one thousand, two hundred and seventeen.

2.2 Representing and describing whole numbers (p.12)

Before we start
Any **three** representations of the number 216 which clearly show the value of each digit, e.g. 200 + 10 + 6.

Questions
1) a) 3262; three thousand, two hundred and sixty-two
   b) 5428; five thousand, four hundred and twenty-eight
   c) 2704; two thousand, seven hundred and four
   d) 6002; six thousand and two

2) a) 200 / two hundred   b) 2000 / two thousand
   c) 20 / twenty   d) 2 / two
   e) 200 / two hundred   f) 2 / two

3) This is an open-ended question. Pupils work in pairs to represent four-digit numbers of their choice with concrete materials, in numerals and in words.

Challenge
She wrote the number 3506. Taking away one hundred and one thousand, and adding a ten, gave her 2416.

2.3 Place value partitioning of numbers with up to five digits (p.14)

Before we start
a) 800, 60 and 7
b) There are five pairs: 34 and 20; 30 + 24; 44 and 10; 40 and 14; 50 and 4. The number 5 does not have a partner.

Questions
1) a) 6348 = 6000 + 300 + 40 + 8
   b) 9564 = 9000 + 500 + 60 + 4
   c) 3715 = 3000 + 700 + 10 + 5
   d) 2285 = 2000 + 200 + 80 + 5
   e) 7491 = 7000 + 400 + 90 + 1

2) a) 3000 | 40 | 2
   b) 7000 | 200 | 90
   c) 5000 | 800 | 9
   d) 2000 | 700 | 60
   e) 3000 | 800
   f) 4000 | 50
   g) 9000 | 3
   h) 2000 | 10 | 1

3) a) Correct   b) Correct
   c) Incorrect. Isla has added 600 to the other numbers instead of 6000. She should write 6194.
   d) Incorrect. Isla has added 900 to the other numbers, instead of 9000. She should write 9049.
   e) Incorrect. Isla has added 70, rather than 7. She should write 7707.
   f) Correct

4) a) 263 matches two hundreds, five tens and 13 ones. It also matches 200 + 50 + 13.
   b) 175 matches 17 tens and five ones. It also matches 170 + 5.
   c) 500 matches four hundreds, nine tens and 10 ones. It also matches 400 + 90 + 10.

d) 408 matches two hundreds, 20 tens and eight ones. It also matches 200 + 200 + 8.

e) 120 matches one hundred, one ten and 10 ones. It also matches 100 + 10 + 10.

f) 297 matches two hundreds and 97 ones. It also matches 200 + 97.

Challenge

a) 20439   b) 62120

2.4 Comparing and ordering numbers in the range 0–10000 (p.18)

Before we start

Isla should look for the number that is worth the least, then the number that is worth the second least and so on. Arranged from smallest to largest the numbers are: 567, 576, 657, 675, 755, 765, 766

Questions

1) a) 2345, 2456, 2718   b) 4132, 4287, 4310
   c) 6014, 6375, 6521

2) a) Smallest to largest: 3627, 3672, 3726, 3762
      Largest to smallest: 3762, 3726, 3672, 3627
   b) Smallest to largest: 8913, 8918, 8925, 8928
      Largest to smallest: 8928, 8925, 8918, 8913
   c) Smallest to largest: 7372, 7374, 7381, 7384
      Largest to smallest: 7384, 7381, 7374, 7372

3) a) Six numbers are possible: 7248, 7284, 7428, 7482, 7824, 7842
   b) Same six numbers written in the order above, i.e. smallest to largest.
   c) Six numbers are possible: 8742, 8724, 8472, 8427, 8274, 8247
   d) Same six numbers written in the order above, i.e. largest to smallest.

Challenge

Nuria is correct because the number 7899 only has seven thousands. The number 8700 has eight thousands.

2.5 Reading and writing decimal fractions (p.20)

Before we start

Amman is correct. The diagram shows $\frac{4}{10}$ because there are 10 equal parts and four of the equal parts are shaded.

Questions

1) a) Three-tenths. Write 0·3. Say **zero point three.**
   b) Seventeen-tenths (or one whole and seven-tenths). Write 1·7. Say **one point seven.**
   c) Nine-tenths. Write 0·9. Say **zero point nine.**

d) Fourteen-tenths (or one whole and four-tenths). Write 1·4. Say **one point four.**

e) Five-tenths. Write 0·5. Say **zero point five.**

f) Eighteen-tenths (or one whole and eight-tenths). Write 1·8. Say **one point eight.**

2) a) 0·6   b) 0·9   c) 2·7
   d) 4·3   e) 5·9

Possible explanations may include:

a) The circle has been divided into 10 equal parts. Each part is one tenth. Six tenths are shaded pink. Four tenths are unshaded.

b) The rectangle has been divided into 10 equal parts. Each part is one tenth. Nine tenths are shaded orange. Only one section is unshaded so nine tenths must be shaded because $\frac{9}{10}$ and $\frac{1}{10} = \frac{10}{10}$ or one whole.

c) There are two whole rectangles and seven tenths of a rectangle.

d) Four whole circles have been shaded and $\frac{3}{10}$ of a circle has been shaded.

e) There are five whole bars of chocolate and $\frac{9}{10}$ of a bar. There are six bars of chocolate. One tenth of the last bar has been eaten so that leaves five whole bars and nine tenths of a bar.

Challenge

The decimal fraction written matches the number of tenths shaded.

2.6 Representing and describing decimal fractions (p.22)

Before we start

a) True   b) True   c) True
d) False   e) False

Questions

1) a) Two different representations of the decimal fraction 1·8 to clearly show one whole item and eight tenths of an item. A satisfactory explanation is provided.

   b) Two different representations of the decimal fraction 2·5 to clearly show two whole items and five tenths of an item. A satisfactory explanation is provided.

   c) Two different representations of the decimal fraction 4·3 to clearly show four whole items and three tenths of an identical item. A satisfactory explanation is provided.

   d) Two different representations of the decimal fraction 5·9 to clearly show five whole items and nine tenths of an identical item. A satisfactory explanation is provided.

2) Frames may be created horizontally or vertically.

Challenge

Isla is correct because the digit after the decimal point represents tenths. $\frac{1}{2}$ is equivalent to $\frac{5}{10}$.

2.7 Comparing and ordering decimal fractions (p.24)

Before we start

The folowing answers are possible:

171 < 317	171 < 731	317 < 731
731 > 317	731 > 171	317 > 171
731 ≠ 317	317 ≠ 731	731 ≠ 171
171 ≠ 731	317 ≠ 171	171 ≠ 317

Questions

1) a) 0·1, 0·3, 0·4, 0·5, 0·6, 0·7 and 0·9 are missing.

   b) 0·5, 0·7, 0·8, 0·9, 1·1, 1·2, 1·3, 1·4 and 1·5 are missing.

   c) 1·8, 1·9, 2·1, 2·2, 2·3, 2·4, 2·5, 2·7 and 2·8 are missing.

2) a) <   b) <   c) >
   d) <   e) =   f) =

3) Fill in the numbers on the number line.

   a)

   Write the correct symbol, <, > or = in each box.

   0·2 < 0·5   0·4 > 0·3   0·7 > 0·5   0·9 < 1·0

   b)

   Write the correct symbol, <, > or = in each box.

   2·5 < 2·9   2·7 > 2·3   2·6 = 2·6   2·8 > 2·4

   c)

   Write the correct symbol, <, > or = in each box.

   4·5 < 5·0   4·4 < 4·8   4·7 > 4·0   4·6 > 1·0

   d)

   Write the correct symbol, <, > or = in each box.

   7·8 > 7·5   7·4 > 7·2   7·7 = 7·7   7·9 > 7·0

   e)

   Write the correct symbol, <, > or = in each box.

   9·3 < 9·5   9·8 > 9·7   9·6 > 9·5   9·9 < 10·0

Challenge

Amman has the least chocolate because he only has one tenth. Finlay has the most chocolate because Isla gave him four tenths which left her with two tenths.

2.8 Recognising the context for negative numbers (p.27)

Before we start

2400, 2800, 2950, 3150, 3500, 3750

Questions

1) Inverness: Nine degrees below zero. –9°C **minus nine degrees Celsius**

Aberdeen: Six degrees below zero. –6°C **minus six degrees Celsius**

Edinburgh: Five degrees below zero. –5°C **minus five degrees Celsius**

Glasgow: One degree below zero. –1°C **minus one degree Celsius**

2) a) Small jelly fish: 1500 m below sea level
   b) Large jelly fish: 500 m below sea level
   c) Orange fish: 3500 m below sea level
   d) Shark: 2000 m below sea level
   e) Starfish: 5500 m below sea level
   f) Yellow fish: 4000 m below sea level
   g) Birds: 2000 m above sea level

3) McIntosh: 5 shots under par   Brown: 3 shots under par

   Smith: 2 shots under par   Gray: 1 shot under par

   Morrison: 1 shot over par   Lyle: 2 shots over par

   Murray: 4 shots over par   Gordon: 6 shots over par

Challenge

Possible scenarios could be: floors in a building, buttons in an elevator, underground car parking levels, bank balances (credit and debit), temperature and pressure gauges, head wind/tail wind differentials, track records or best lap (+ or – seconds/minutes).

3 Number – addition and subtraction

3.1 Mental addition and subtraction (p.30)

Before we start

   a) 78 – appropriate strategies include round and adjust (80 – 2); partitioning (60 + 18).

   b) 545 – appropriate strategies include counting on from 537 (+ 3 is 540 + 5 is 545): add 10, then subtract 2.

   c) 60 – appropriate strategies include doubling/halving (60 + 60 = 120); using known facts and place value (12 – 6 = 6 so 120 – 60 = 60).

   d) 249 – appropriate strategies include subtract 10, then add 1.

Questions

1) a) 38   b) 82   c) 116   d) 247
   e) 637   f) 281   g) 155   h) 295
   i) 446

2) a) 95   b) 70   c) 45   d) 130
   e) 125   f) 295   g) 850   h) 474
   i) 384   j) 470

3) Any **five** of the following addition questions and answers with a correct explanation of the mental strategy used:

125 + 19 = 144	19 + 125 = 144
125 + 21 = 146	21 + 125 = 146
125 + 29 = 154	29 + 125 = 154
125 + 131 = 256	131 + 125 = 256
54 + 19 = 73	19 + 54 = 73
54 + 21 = 75	21 + 54 = 75
54 + 29 = 83	29 + 54 = 83
54 + 131 = 185	131 + 54 = 185
66 + 19 = 85	19 + 66 = 85
66 + 21 = 87	21 + 66 = 87
66 + 29 = 95	29 + 66 = 95
66 + 131 = 197	131 + 66 = 197
78 + 19 = 97	19 + 78 = 97
78 + 21 = 99	21 + 78 = 99
78 + 29 = 107	29 + 78 = 107
78 + 131 = 209	131 + 78 = 209
64 + 19 = 83	19 + 64 = 83
64 + 21 = 85	21 + 64 = 85
64 + 29 = 93	29 + 64 = 93
64 + 131 = 195	131 + 64 = 195

Any **five** of the following subtraction questions and answers with a correct explanation of the mental strategy used:

125 – 19 = 106	125 – 21 = 104
125 – 29 = 96	131 – 125 = 6
54 – 19 = 35	54 – 21 = 33
54 – 29 = 25	131 – 54 = 77
66 – 19 = 47	66 – 21 = 45
66 – 29 = 37	131 – 66 = 65
78 – 19 = 59	78 – 21 = 57
78 – 29 = 49	131 – 78 = 53
64 – 19 = 45	64 – 21 = 43
64 – 29 = 35	131 – 64 = 67

Challenge

a) 52 + 31 = 83   b) 56 – 39 = 17

3.2 Adding and subtracting 1, 10, 100 and 1000 (p.32)

Before we start

The number on Amman's mini whiteboard is 390.

Questions

1) a) 1127, 1128, 1129, 1130, 1131
    b) 4696, 4697, 4698, 4699, 4700
    c) 8999, 9000, 9001, 9002, 9003
    d) 2872, 2871, 2870, 2869, 2868
    e) 5200, 5199, 5198, 5197, 5196
    f) 7003, 7002, 7001, 7000, 6999

2) a) 4950, 6781, 1600, 4000
    b) 6019, 8299, 2092, 4999

3) a) Starting number 2500: 2600, 2700, 2800, 2900
       Starting number 3426: 3526, 3626, 3726, 3826
       Starting number 1900: 2000, 2100, 2200, 2300
       Starting number 6820: 6920, 7020, 7120, 7220
    b) Starting number 9999: 8999, 7999, 6999, 5999
       Starting number 7081: 6081, 5081, 4081, 3081
       Starting number 4200: 3200, 2200, 1200, 200
       Starting number 4001: 3001, 2001, 1001, 1

Challenge

1) a) 3791   b) 3800   c) 3890   d) 4790
    e) 3789   f) 3780   g) 3690   h) 2790

3.3 Adding and subtracting multiples of 100 (p.34)

Before we start

a) 180 and 120   b) 50, 100 and 150
c) 70, 140, 60 and 30 or 150 + 70 + 30 + 50

Questions

1) a) Amman scored the same as Finlay.
    b) Nuria scored 2600.
    c) 2600, 1800, 1400
    d) Any **three** of the following:
       500 + 500 + 500 + 500 = 2000
       800 + 200 + 700 + 300 = 2000
       900 + 100 + 500 + 500 = 2000
       900 + 100 + 700 + 300 = 2000
       500 + 500 + 700 + 300 = 2000

2) a) First row: 21   Second row: 90 and 210
       Third row: 300 and 2100
    b) First row: 11   Second row: 50 and 110
       Third row: 1000, 800 and 1100
    c) First row: 36   Second row: 60 and 360
       Third row: 500, 900 and 3600

Challenge

a) 3500   b) 2100   c) 1400

3.4 Adding and subtracting by making 10s or 100s (p.36)

Before we start

a) Correct

b) Incorrect. Finlay should have written 19 + 81 = 100 or 29 + 71 = 100

c) Incorrect. Finlay should have written 55 + 45 = 100 or 45 + 55 = 100

d) Incorrect. Finlay should have written 100 – 27 = 73 or 100 – 17 = 83

Questions

1) a) 787   b) 500   c) 726
    d) 628   e) 831   f) 715

2) a) 376   b) 262   c) 381

3) a) 126   b) 311   c) 719

Challenge

The digits in bold print complete the number sentences:
4**33** + 67 = 500   600 – 24**5** = 355

3.5 Adding and subtracting multiples of 1000 (p.38)

Before we start

First row: 250

Second row: any two multiples of 10 that total 400

Questions

1) a) The missing numbers are: 5000, 100, 70 and 5.
    3000 + 5175 = 8175   8175 – 5175 = 3000

b) The missing numbers are: 7, 90, 700 and 2000.
    1203 + 2797 = 4000   4000 – 2797 = 1203

c) The missing numbers are: 4000, 900, 90 and 9.
    5000 + 4999 = 9999   9999 – 4999 = 5000

2) a) 6532   b) 3778   c) 1570
    d) 5305   e) 1445   f) 1439

Challenge

a) 1700

b) Any **two** four-digit numbers that total 3700.

c) Any **two** four-digit numbers that total 2585.

3.6 Adding and subtracting multiples of 10 and 100 (p.40)

Before we start

a) 417 + $\boxed{4}\boxed{0}\boxed{0}$ = 817 Missing digits are 4, 0 and 0

b) 673 + $\boxed{2}$ 0 0 = 8 $\boxed{7}$ $\boxed{3}$ Missing digits are 2, 7 and 3

c) 349 – $\boxed{3}\boxed{0}\boxed{0}$ = 49 Missing digits are 3, 0 and 0

d) $\boxed{6}$ 21 – 121 = 5 $\boxed{0}\boxed{0}$ Missing digits are 6, 0 and 0

Questions

1) a) 7504   b) 6625   c) 6751
    d) 6077   e) 6034   f) 3355
    g) 3229   h) 5608   i) 9453

2) a) 9683   b) 7898   c) 6785
    d) 8984   e) 2327   f) 4285
    g) 2110   h) 3529   i) 1122

3) a) 2335 + $\boxed{600}$ = 2935   b) 5922 + $\boxed{40}$ = 5962
    c) 1371 + $\boxed{320}$ = 1691   d) 3587 – $\boxed{70}$ = 3517
    e) 8529 – $\boxed{200}$ = 8329   f) 6745 – $\boxed{140}$ = 6605

Challenge

Finlay is incorrect because he has subtracted 200 from 800 instead of 800 from 200. His strategy does not work in this instance because he does not have enough hundreds. The correct answer to Isla's question is 1443.

3.7 Solving word problems (p.42)

Before we start

The missing numbers are 207 and 610.

Questions

1) 7000

2) a) 4026   b) 326

3) 1760

4) 114

5) a) 541   b) 3747

Challenge

Any word problem that matches the bar model given.

	876	
137	239	?

3.8 Representing word problems in different ways (p.44)

Before we start

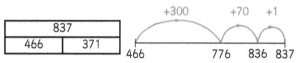

1) a)

Word problem

In a TV dancing contest 5356 people voted for Couple Number One, 3206 people voted for Couple Number Two and 6864 voted for Couple Number Three. Find the difference between the highest and lowest scores.

Bar model

6864	
3206	**3658**

*3206 and 3658 are interchangeable

Empty number line

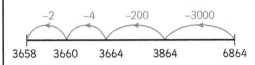

−2	−4	−200	−3000

3658 3660 3664 3864 6864

*Alternative methods are possible

Number sentence

6864 − 3206 = 3658

*Alternative number sentences are possible. Answer should reflect the method used and clearly illustrate that the difference between 6864 and 3206 is 3658.

1) b)

Word problem

Isla's mum delivered 2768 leaflets between Monday and Thursday. How many leaflets does she still need to deliver if she started with 3200?

Bar model

3200	
2768	**432**

*2768 and 432 are interchangeable

Empty number line

An empty number line which shows:

- Counting on from 2768 up to 3200 to reach the answer 432

OR

- Counting back from 3200 down to 2768 to reach the answer 432

Number sentence

3200 − 2768 = 432

*Alternative number sentences are possible. Answer should reflect the method used and clearly illustrate that the difference between 3200 and 2768 is 432.

1) c)

Word problem

Finaly and his mum are at a football match. There are 868 spectators in the north stand. The stand can hold 2000 people. How many empty seats are there?

Bar model

2000	
868	**1132**

*868 and 1132 are interchangeable

Empty number line

An empty number line which shows:

- Counting on from 868 up to 2000 to reach the answer 1132

OR

- Counting back from 2000 down to 868 to reach the answer 1132

Number sentence

2000 − 868 = 1132

*Alternative number sentences are possible. Answer should reflect the method used and clearly illustrate that the difference between 2000 and 868 is 1132.

2) a) Nuria is correct. Satisfactory explanations include:

   It is subtraction because:

   you are finding the difference.

   you are comparing the journeys to find how much further away London is.

   b) Problem represented as a bar model, empty number line and number sentence, all of which clearly show that the difference between 1000 and 339 is 661.

Challenge

Many solutions are possible.

3.9 Using non-standard place value partitioning (p.46)

Before we start

68 can be 6 packs of 10 and 8 single candy canes.
60 + 8 = 68

68 can be 4 packs of 10 and 28 single candy canes.
40 + 28 = 68

68 can be 3 packs of 10 and 38 single candy canes.
30 + 38 = 68

68 can be 2 packs of 10 and 48 single candy canes.
20 + 48 = 68

68 can be 1 pack of 10 and 58 single candy canes.
10 + 58 = 68

68 can be 68 single candy canes. 0 + 68 = 68

Questions

1) a) 1 box of 100, 3 packs of 10 and 5 single candy canes

   b) 1 box of 100, 2 packs of 10 and 15 single candy canes

   c) 1 box of 100, 1 pack of 10 and 25 single candy canes

   d) 1 box of 100 and 35 single candy canes

2) a) 2 hundreds, 18 tens and 2 ones = 200 + 180 + 2 = 382

   b) 3 hundreds, 12 tens and 7 ones = 300 + 120 + 7 = 427

   c) 6 hundreds, 11 tens and 4 ones = 600 + 110 + 4 = 714

Challenge

a) 756 = 700 + 30 + 26   756 = 700 + 20 + 36
   756 = 700 + 10 + 46   756 = 700 + 56

b) 756 = 600 + 140 + 16

c) Answers will vary.

3.10 Adding using a semi-formal written method (p.48)

Before we start

600 + 150 + 6 = 756   500 + 250 + 6 = 756
400 + 350 + 6 = 756   300 + 450 + 6 = 756
200 + 550 + 6 = 756   100 + 650 + 6 = 756

Questions

1) a) 783   b) 771   c) 863
   d) 562   e) 955   f) 466
   g) 729   h) 820   i) 840
   j) 754   k) 1032   l) 1332

2) a) 900   b) 1663   c) 920

3) a) 5368 + [500] − [30] = 5838

   b) 6189 − [40] + [300] = 6449

   c) 2035 − [1000] − [20] + [700] = 1715

Challenge

a) 200 + 150 + 3 (palindrome 353)

b) 400 + 560 + 9 (palindrome 969)

c) 500 + 320 + 8 (palindrome 828)

d) 300 + 210 + 5 (palindrome 515)

3.11 Adding three-digit numbers using standard algorithms (p.50)

Before we start

1 hundred, 4 tens, 9 ones

14 tens, 9 ones

1 hundred, 49 ones

Questions

1) a) 929   b) 770   c) 827
   d) 1358   e) 762   f) 905
   g) 921   h) 1352   i) 823
   j) 1628   k) 1555   l) 934

2) a) 1050   b) 1944   c) 1690   d) 1085

3) Answers will vary.

Challenge

247 + 382 = 629 and 508 + 629 = 1137

3.12 Representing and solving word problems (p.52)

Before we start

No, Isla does not have enough lemonade. 330 + 545 = 875 so she needs 125 ml more.

Questions

1) 167 miles

2) 189 people

3) 45 miles

4) a) 1615 people   b) 379 children

5) 832

6) 416 keyrings

Challenge

a) 239 + 677 = 916   677 + 239 = 916

    916 − 677 = 239   916 − 239 = 677

b) Appropriate word problem that matches one of the above number sentences.

4 Number – multiplication and division

4.1 Recalling multiplication and division facts for 2, 5 and 10 (p.54)

Before we start

2 × 10 = 20

Questions

1) a) 12   b) 30   c) 80

    d) 18   e) 20   f) 30

2) a) 12, 30, 60   b) 8, 20, 40   c) 20, 50, 100

    d) 18, 45, 90

3) a) 7   b) 6   c) 80

    d) 1   e) 9   f) 10

    g) 18   h) 8   i) 4

    j) 4   k) 7   l) 5

Challenge

Answers will vary.

4.2 Recalling multiplication and division facts for 3 (p.56)

Before we start

Array for 5 × 3:

⚫ ⚫ ⚫ ⚫ ⚫
⚫ ⚫ ⚫ ⚫ ⚫
⚫ ⚫ ⚫ ⚫ ⚫

To change the array to show 6 × 3, you would add another column of three to the end of the diagram.

Questions

1) a) 6   b) 4   c) 27

    d) 5   e) 3   f) 30

    g) 2   h) 8   i) 9

    j) 7   k) 10   l) 9

2) a) 3   b) 2   c) 5

    d) 7   e) 12   f) 9

3) a) 9 ÷ 3 = 3

    b) 6 ÷ 3 = 2 or 6 ÷ 2 = 3

    c) 15 ÷ 3 = 5 or 15 ÷ 5 = 3

    d) 21 ÷ 3 = 7 or 21 ÷ 7 = 3

    e) 12 ÷ 3 = 4 or 12 ÷ 4 = 3

    f) 27 ÷ 3 = 9 or 27 ÷ 9 = 3

Challenge

Answers will vary.

4.3 Solving multiplication problems (p.58)

Before we start

20 muffins

Questions

1) a) 35   b) 14   c) 49   d) 35 + 14 = 49

2) (5 × 8) + (2 × 8) = 56

3) a) (5 × 4) + (2 × 4) = 28

    b) (5 × 6) + (1 × 6) = 36

    c) (5 × 6) + (2 × 6) = 42

    d) (2 × 6) + (2 × 6) = 24

    e) (10 × 7) + (2 × 7) = 84

    f) (10 × 6) + (5 × 6) = 90

Challenge

(10 × 6) + (10 × 6) + (5 × 6) = 150

4.4 Multiplying whole numbers by 10 and 100 (p.60)

Before we start

6 × 4 = 24

Questions

1) a) 10 × 3   b) 10 × 4   c) 10 × 6

2) a) 7 × 100 = 700   b) 9 × 100 = 900

    c) 14 × 100 = 1400   d) 24 × 100 = 2400

    e) 56 × 100 = 5600

3) a) 600 cm   b) 900 cm   c) 1200 cm

    d) 1800 cm   e) 2100 cm

Challenge

Amman started with the number 6.

Multiplying by ten, and by then ten again is the same as multiplying by 100.

4.5 Dividing whole numbers by 10 and 100 (p.62)

Before we start

20 ÷ 5 = 4

Questions

1) a) $30 \div 10 = 3$ b) $50 \div 10 = 5$ c) $20 \div 10 = 2$
   d) $10 \div 10 = 1$ e) $80 \div 10 = 8$ f) $20 \div 10 = 2$

2) a) 30 divided by 10 = 3   b) 100 divided by 10 = 10
   c) 50 divided by 10 = 5   d) 120 divided by 10 = 12

3) a) $7 \times 100 = 700$ 700 divided by 100 = 7
   b) $100 \times 10 = 1000$ 1000 divided by 10 = 10
   c) $9 \times 100 = 900$ 900 divided by 100 = 9
   d) $15 \times 100 = 1500$ 1500 divided by 100 = 15
   e) $24 \times 100 = 2400$ 2400 divided by 100 = 24

Challenge

$200\,cm = 2\,m$ $8500\,cm = 85\,m$ $3100\,cm = 31\,m$
$1400\,cm = 14\,m$ $2300\,cm = 23\,m$ $9900\,cm = 99\,m$

4.6 Solving multiplication problems by partitioning (p.64)

Before we start

12

Questions

1) $21 \times 3 = (20 \times 3) + (1 \times 3) = 60 + 3 = 63$
2) $28 \times 5 = (20 \times 5) + (8 \times 5) = 100 + 40 = 140$
3) $33 \times 4 = (30 \times 4) + (3 \times 4) = 120 + 12 = 132$

Challenge

Answers will vary.

4.7 Solving multiplication and division problems (p.66)

Before we start

a) False – 10 is half of 20
b) False – double 6 is 12
c) True
d) True
e) False – 7 doubled is 14
f) True

Questions

1) a) 8, double 40 = 80   b) 12, double 60 = 120
   c) 24, double 120 = 240   d) 30, double 150 = 300

2) a) Nuria has baked 30 samosas. 15 + 15 = 30, $15 \times 2 = 30$
   b) Isla's brother weighs 40 kg. 20 + 20 = 40, $20 \times 2 = 40$
   c) Nuria thinks there are 600 cars in the car park. 300 + 300 = 600, $300 \times 2 = 600$

3) a) 3; 60 = 30   b) 7; 140 = 70   c) 10; 200 = 100

Challenge

Option 1: $100 \times 30 = £3000$
Option 2: Money doubles each day, so day 1: £1, day 2: £2, day 3: £4, day 4: £8... and so on.
If you wanted to win more money, you would choose option 2, because by day 12 you would have won more than £3000 – the total amount you would win if you chose option 1.

4.8 Recalling multiplication and division facts for 4 (p.68)

Before we start

a) incorrect: $2 \times 7 = 14$ b) incorrect: $8 \times 2 = 16$
c) correct   d) incorrect: double 6 is 12
e) correct   f) incorrect: half of 12 is 6

Questions

1) Answers will vary.

2) a) There are nine 4s in 36   b) Five lots of 4 are 20
   c) Three groups of 4 make 12   d) Seven 4s are 28

3) a) 3   b) 5   c) 40   d) 16
   e) 28   f) 4   g) 9   h) 2

Challenge

$56 \div 4 = £14$

4.9 Solving division problems (p.70)

Before we start

$3 \times 5 = 15$

Questions

1) a) Nine bracelets   b) 13 rows   c) Six packets

2) Six sweets

3) Finlay spent £3 each day. It took him eight days to spend £24.
   Nuria spent £4 each day. It took her six days to spend £24.
   Amman spent £6 each day. It took him four days to spend £24.

Challenge

1, 2, 3, 4, 5, 6, 10, 12, 15, 20, 30, 60

4.10 Multiplying by changing the order of factors (p.72)

Before we start

a) 4   b) 3   c) 40   d) 5   e) 10   f) 9   g) 6   h) 14

Questions

1) a) $2 \times 6 = 12$ b) $4 \times 3 = 12$
   c) $7 \times 5 = 35$ d) $8 \times 10 = 80$

2) a) True   b) False

    c) False   d) True

3) a) $3 \times 7 = 7 \times 3$ b) $4 \times 8 = 8 \times 4$

    c) $9 \times 2 = 2 \times 9$ d) $4 \times 6 = 6 \times 4$

    e) $5 \times 9 = 9 \times 5$ f) $20 \times 2 = 2 \times 20$

4) Answers will vary

Challenge

Answers will vary.

4.11 Solving multiplication problems (p.74)

Before we start

$3 \times 2 = 6$; $3 \times 4 = (3 \times 2) + (3 \times 2) = 12$;
$3 \times 12 = (3 \times 2) + (3 \times 10) = 36$

Questions

1) a) $3 \times 7 = (2 \times 7) + (1 \times 7) = 21$

    b) $6 \times 7 = (3 \times 7) + (3 \times 7) = 42$

    c) $4 \times 7 = (2 \times 7) + (2 \times 7) = 28$

    d) $8 \times 7 = (4 \times 7) + (4 \times 7) = 56$

    e) $12 \times 7 = (6 \times 7) + (6 \times 7) = 84$

2) a) $3 \times 10 = 30$; double it = 60

    b) $5 \times 6 = 30$; double it = 60

    c) $4 \times 8 = 32$; double it = 64

    d) $4 \times 6 = 24$; double it = 48

3) a) 10   b) 20   c) 40   d) 44

Challenge

Answers will vary.

4.12 Using known facts and halving to solve division problems (p.76)

Before we start

    a) 16, 16 halved is 8

    b) 10, 10 halved is 5

    c) 18, 18 halved is 9

    d) 12, 12 halved is 6

Questions

1) a) $2 \times 8 = 16$ $16 \div 2 = 8$ $16 \div 4 = 4$

    b) $2 \times 4 = 8$ $8 \div 2 = 4$ $8 \div 4 = 2$

    c) $2 \times 6 = 12$ $12 \div 2 = 6$ $12 \div 4 = 3$

    d) $2 \times 12 = 24$ $24 \div 2 = 12$ $24 \div 4 = 6$

2) a) 5   b) 10   c) 5

    d) 9   e) 7   f) 6

3) a) 32 shared between four bags would be eight marbles each.

    b) 32 shared between eight would be four marbles each.

    c) 32 shared between 16 bags would be two marbles each.

4) 6 rows

Challenge

Answers will vary.

5 Multiples, factors and primes

5.1 Recognising multiples and factors (p.78)

Before we start

$20 \div 4 = 5$

Questions

1) a) 3   b) 3   c) 4   d) 5

    e) $1 \times 6 = 6$ or $2 \times 3 = 6$

    f) $1 \times 30 = 30$, $2 \times 15 = 30$, $3 \times 10 = 30$, $5 \times 6 = 30$

    g) $1 \times 12 = 12$, $2 \times 6 = 12$, $3 \times 4 = 12$

    h) $1 \times 20 = 20$, $2 \times 10 = 20$, $4 \times 5 = 20$

2) a) 3, 6, 9, 12, 15   b) 5, 10, 15, 20, 25

    c) 4, 8, 12, 16, 20   d) 10, 20, 30, 40, 50

3) a) True   b) False   c) False   d) True

    e) True   f) False   g) True   h) False

Challenge

Answers will vary.

6 Fractions, decimal fractions and percentages

6.1 Comparing fractions (p.80)

Before we start

Check students' foldings.

Questions

1) a) One twelfth, one tenth, one eighth, one sixth, one quarter, one half

    b)

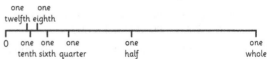

2) a) Answers will vary.

b)

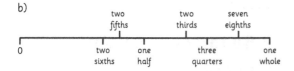

Challenge
Answers will vary.

6.2 Identifying equivalent fractions pictorially (p.83)

Before we start
Check students' drawings.

Questions
1) a – three sixths, d – six twelfths, g – four eighths
2) a) b – three ninths, e – two sixths, i – four twelfths

b) Answers will vary.
3) a) c – 15 twentieths, f – six eighths, h – nine twelfths

b) Answers will vary.

Challenge
Answers will vary, e.g. one third = two sixths

6.3 Identifying and creating equivalent fractions (p.86)

Before we start
Answers will vary, e.g. two quarters, three sixths, four eighths.

Questions
1) a) Two eighths   b) Three twelfths c) Five twentieths
2) Answers will vary, e.g. six eighths, nine twelfths, 15 twentieths
3) a) 25 parts

b) i) 25 hundredths   ii) 50 hundredths

iii) 75 hundredths   iv) 100 hundredths

Challenge
Check students' foldings.

6.4 Simplifying fractions (p.88)

Before we start
Answers will vary, e.g. cut it into eighths and give each child two slices.

Questions
1) one half = 10 twentieths   two thirds = six ninths

three fifths = six tenths   five eighths = 10 sixteenths

2) four sixths = two thirds

six eighths = three quarters

five twentieths = one quarter

eight tenths = four fifths

3) a)

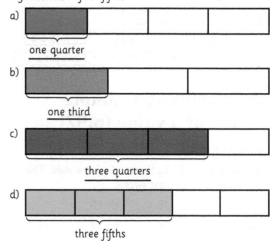

one quarter

b)

one third

c)

three quarters

d)

three fifths

Challenge
Answers will vary.

6.5 Writing decimal equivalents to tenths (p.91)

Before we start
Answers will vary, e.g. No, I don't agree with Finlay. One half equals five tenths and one fifth equals two tenths, so splitting them both into 10 equal pieces would not be correct.

Questions
1) a) three tenths = 0·3   b) six tenths = 0·6

c) eight tenths = 0·8   d) 10 tenths = 1 or 1·0
2) a) One half = five tenths = 0·5
3) Two fifths = four tenths = 0·4
4) a) three whole and six tenths = 36 tenths = 3·6

b) two whole and five tenths = 25 tenths = 2·5

Challenge
a) three whole and five tenths = 3·5 pizzas

two whole and eight tenths = 2·8 pizzas

6.6 Comparing numbers with one decimal place (p.94)

Before we start
Answers will vary, e.g. 9.1, 9.3, 9.4, 9.6, 9.9.

Questions
1) b) 0·2 is smaller than 0·8

c) 1·4 is larger than 0·9

d) 2·4 is smaller than 2·5

2)

| 0·4 | 0·5 | | 0·8 | 1·0 | | 1·4 | 1·5 | 1·7 | 1·8 | 2·0 | | 2·2 |

3)   Lauren Elie Finn Marissa Hamza Mark Stuart Ava

Challenge

2·8	2·9	3·0	3·1	**3·2**
4·6	4·7	4·8	4·9	5·0	5·1	5·2	**5·3**
19·7	19·8	19·9	20·0	20·1	20·2	20·3	20·4	**20·5**

6.7 Calculating a simple fraction of a value (p.97)

Before we start

Nuria should divide 24 by three to find one third. This means she should give Finlay eight sweets.

1)   a) 12   b) 18   c) 10

2)   a) 30   b) 28   c) 12   d) 21

3)   a) 20 girls   b) 42 children   c) £42

Questions

1)   a) 12   b) 18   c) 10

2)   Check students' bar models: a) 30   b) 20   c) 12   d) 21

3)   a) 20   b) 42   c) £42

Challenge

Answers will vary, e.g. Nuria is wrong as the bar shows quarters not thirds. She should count the total number of sections first.

7 Money

7.1 Writing amounts using decimal notation (p.100)

Before we start

Six pounds and four pence	604p	£6·04
Nine pounds and thirty-six pence	936p	£9·36
Eight pounds and twenty-four pence	824p	£8·24
Nine pounds and forty pence	940p	£9·40
Twenty-three pence	23p	£0·23

Questions

1)   a) £5·23   b) £0·94   c) £12·08
     d) £30·00   e) £1·00   f) £0·60

2)   a) Yes, since 12 × 45 pence is £5·40, which is less than £6.
     b) 60p

3)   13 days

Challenge

Answers will vary.

7.2 Budgeting (p.102)

Before we start

Total spend £15·13
Change £4·87

Questions

1)   11 weeks

2)   c) Crop jeans and vest top

3)   a) Train   b) £21·25   c) £3·75

Challenge

   a) Three friends, remember Isla has to pay for herself
   b) £70
   c) Yes, she will have £5 change.

7.3 Saving money (p.105)

Before we start

£9·05

Questions

1)   20p

2)   a) Forthside book of 10 tickets   b) 50p

3)   a) £1795
     b) Forever Travel and Roland Air
     c) £80

Challenge

   a) Supersavers Superstore   b) 31p   c) £9·17

7.4 Profit and loss (p.108)

Before we start

Answers will vary, e.g. fidget spinner, pencil case, football and football shirt

Questions

1)   £90

2)   £450

3)   £30 loss

4)   £9875

Challenge

   a) £164
   b) £191

8 Time

8.1 Telling the time to the minute – 12-hour clock (p.110)

Before we start

![clock showing 3:15]	3:15 or 15:15	Fifteen minutes past three
![clock showing 12:35]	![digital clock 12:35] 12:35	Twenty-five minutes to one
![clock showing 3:55]	3:55 or 15:55	Five minutes to four
![alarm clock showing 10:10]	10:10 or 22:10	Ten minutes past ten

Questions

1) b) Fifty-four minutes past three/six minutes to four

  c) Twenty-two minutes past twelve/thirty-eight minutes to one

  d) Fifty-seven minutes past eight/three minutes to nine

  e) Twenty-three minutes past two/thirty-seven minutes to three

  f) Forty-six minutes past four/fourteen minutes to five

  g) Twenty-four minutes past eleven/thirty six minutes to twelve

  h) Fifty-six minutes past ten/four minutes to eleven

2) a) Twenty-one minutes past one

   b) Twenty-nine minutes to nine

   c) Six minutes past 12

   d) Nine minutes to 11

3) a) Forty-four minutes past eight in the morning

  b) Eighteen minutes past four in the morning

  c) Forty-nine minutes past seven in the evening

  d) Twenty-three minutes past two in the afternoon

4) a) ![clock] 7 43 PM

   Forty-three minutes past seven/ seventeen minutes to eight in the evening

   b) 11 19 AM ![clock]

   Nineteen minutes past eleven in the morning

   c) ![clock] 1 11 AM

   Eleven minutes past one in the morning

   d) 9 57 AM ![clock]

   Fifty-seven minutes past nine/three minutes to ten in the morning

Challenge
Answers will vary.

8.2 Converting between 12-h and 24-h time (p.114)

Before we start
24 hours = one day; 2:00 am = 0200; 3:00 am = 0300; 4:00 pm = 1600; 5:00 pm = 1700; one year = 365 days; seven days = one week

Questions

1) a) 2135   b) 0330   c) 0445

2) a) 6:40 pm   b) 3:40 am   c) 9:22 pm

3) a) 8:44 am / 0844

  b) 7:49 pm / 1949

  c) 5:49 pm / 1749

  d) 3:13 am / 0313

Challenge

12-hour time	24-hour time
2:15 am	0215
6:46 pm	1846
9:04 pm	2104
Noon	1200
4:28 am	0428
1:14 pm	1314

8.3 Converting minute intervals to fractions of an hour (p.116)

Before we start
60 minutes / one hour

Questions

1) a)   b)   c)   d)

2) a) Quarter to eleven
    b) Half past five
    c) Quarter past eight
    d) Quarter to ten

Challenge

1) a) Quarter to ten / 9:45 am

    b) Half past one / 1:30 pm

    c) Quarter past ten / 10:15 am

2) One hour

8.4 Calculating time intervals or durations (p.118)

Before we start
One hour 35 minutes

Questions

1) 7:00 pm → 7:30 pm → 8:00 pm → 8:30 pm → 9:00 pm

2) a) One hour 30 minutes
    b) Two hours 30 minutes
    c) One hour 40 minutes
    d) Two hours
    e) Three hours
    f) Four hours 30 minutes

3) a) 10 minutes   b) 50 minutes   c) 0900

Challenge
2·20 am

8.5 Speed, time and distance calculations (p.120)

Before we start
9:10 am

Questions

1) a) 135 miles   b) 20 miles
2) a) 2160 miles   b) 840 miles
3) Three hours / 195 miles

Challenge
290 miles

9 Measurement

9.1 Using familiar objects to estimate length, mass, area and capacity (p.122)

Before we start
length – e.g. cm; area – e.g. cm^2; mass – e.g. g;
capacity – e.g. ml

Questions

1) Sledge = 1 m; golf cart = 3 m; car = 4 m; bus = 6 m;
    aeroplane = 8 m

2) Kitchen = 8 m^2; hallway = 28 m^2; bedroom 1 = 24 m^2;
    bedroom 2 = 16 m^2; lounge = 40 m^2

Challenge
Answers will vary.

9.2 Estimating and measuring length (p.126)

Before we start
Answers will vary.

Questions

1) a) 7 cm or 70 mm   b) 12 cm or 120 mm
    c) 10 cm or 100 mm   d) 9 cm or 90 mm
    e) 5 cm or 50 mm   f) 13 cm or 130 mm

2) a) 6 cm or 60 mm   b) 3 cm or 30 mm
    c) 9 cm or 90 mm   d) 10 cm or 100 mm
    e) 13 cm or 130 mm   f) 14 cm or 140 mm

3) Answers will vary.

Challenge
Answers will vary.

9.3 Estimating and measuring mass (p.129)

Before we start
Answers will vary.

Questions
1) a) 100 g   b) 250 g   c) 600 g   d) 900 g
   e) 2 kg   f) 2400 g   g) 2750 g   h) 3300 g
   i) 3·5 kg

2) Answers will vary.

Challenge
Answers will vary.

9.4 Converting units of length (p.132)

Before we start
2·2 cm (22 mm), 3·8 cm (38 mm), 5·5 cm (55 mm)

Questions
1) Horse = 182 cm; Finlay = 122 cm; Nuria = 109 cm;
   cat = 47 cm; cow = 145 cm; elephant = 339 cm;
   giraffe = 450 cm

2) Ant = 28 mm; butterfly = 40 mm; frog = 75 mm;
   mouse = 102 mm; caterpillar = 61 mm; spider = 37 mm;
   fish = 99 mm; guinea pig = 120 mm

Challenge
Answers will vary.

9.5 Calculating the perimeter of simple shapes (p.135)

Before we start
3 cm (30 mm); 5 cm (50 mm); 6·5 cm (65 mm)

Questions
1) a) 16 cm   b) 14 cm   c) 13 cm   d) 20 cm

2)

Shape	No. of measurements	Perimeter
rectangle	2	12 cm
square	1	12 cm
isosceles triangle	2	13·5 cm
parallelogram	2	10 cm
hexagon	1	9 cm

Challenge
Answers will vary.

9.6 Finding the area of regular shapes in square cm/m (p.138)

Before we start
24 soldiers
20 soldiers, six ways: 1 × 20, 2 × 10, 4 × 5, 5 × 4,
10 × 2, 20 × 1

Questions
1) a) 14 cm²   b) 25 cm²   c) 30 cm²   d) 24 cm²

2) a) 24 cm²   b) 48 cm²   c) 16 cm²

Challenge
Answers will vary.

9.7 Finding the volume of cubes and cuboids by counting cubes (p.141)

Before we start
6 cm³ and 12 cm³

Questions
1) a) 19 ml   b) 22 ml   c) 750 ml
   d) 400 ml   e) 700 ml   f) 1700 ml

2) Answers will vary.

Challenge
Answers will vary.

9.8 Estimating and measuring capacity (p.144)

Before we start
28 cm²   24 cm²

Questions
1) a) 8 cm³   b) 12 cm³
   c) 16 cm³   d) 18 cm³

2) a) 2 × 2 × 2
   b) 4 × 1 × 3
   c) 4 × 2 × 2
   d) 3 × 2 × 3

3) a) 2 cm × 3 cm × 2 cm = 12 cm³
   b) 2 cm × 5 cm × 2 cm = 20 cm³
   c) 3 cm × 3 cm × 2 cm = 18 cm³
   d) 3 cm × 2 cm × 4 cm = 24 cm³
   e) 3 cm × 3 cm × 3 cm = 27 cm³

Challenge
a) Nuria
b) Amman: three shapes, Isla: one shape, Finlay: two
   shapes, Nuria: four shapes

10 Mathematics, its impact on the world, past, present and future

10.1 Mathematical inventions and different number systems (p.146)

Before we start

Answers will vary.
Blaise Pascal: Pascal's Triangle
Leonardo Fibonacci: Fibonacci Sequence

Questions

1) a) XXII   b) XXXI   c) XXIX
   d) XLV   e) XCIII   f) CXXVI
   g) DLV   h) MMXVIII

2) Answers will vary.

3) a) IX   b) XXI   c) CCCL
   d) XLVIII

Challenge

Answers will vary, but will include:

a) John Napier of Merchiston born 1550, died 4 April 1617, nicknamed Marvellous Merchiston

b) from Edinburgh

c) most famous for logarithms.

11 Patterns and relationships

11.1 Exploring and extending number sequences (p.148)

Before we start

The next shape is

The pattern is two stars then a green circle, and then the pattern repeats itself.

Questions

1) a) 🌙   b) ⬜   c) M

2) a) 34, 55 The rule is add the previous two numbers together.

   b) 49, 57 The rule is add eight.

   c) 57, 50 The rule is subtract seven.

Challenge

Answers will vary.

12 Expressions and equations

12.1 Solving simple equations using known number facts (p.150)

Before we start

a) True   b) False
c) False   d) False

Questions

1) a) 5   b) –   c) 77
2) a) 3   b) 6   c) 3
3) a) 5   b) 6   c) 35

Challenge

a) $6 \times \mathbf{5} = 98 - 68$ b) $545 - 425 = \mathbf{10} \times 12$
c) $675 \div \mathbf{3} = 5 \times 45$ d) $650 \div 10 = \mathbf{22} + 43$

13 2D shapes and 3D objects

13.1 Drawing 2D shapes (p.152)

Before we start

There are two right angles in the shape, so Isla is correct.

Questions

1) Check students' drawings.

2) Check students' drawings.

3) a) Check students' drawings and ensure first square is 4 cm and each subsequent square is 2 cm bigger.

   b) Check students' drawings.

Challenge

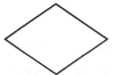

Where the sides are 5 cm.

13.2 Naming and sorting 2D shapes (p.154)

Before we start

Nuria is right: the shape is an irregular pentagon. It has five straight sides and five vertices.

Questions

1)

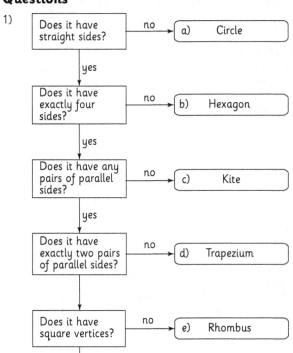

2) a) Shape A   b) Shape B
    c) Shape E

Challenge

The shape is neither a kite nor a trapezium. The adjacent sides are not of equal length so it cannot be a kite. It has no pairs of parallel sides so it is not a trapezium. It is an irregular quadrilateral.

13.3 Drawing 2D shapes and 3D objects (p.157)

Before we start

An appropriate justification for any of the shapes being the odd one out.

Questions

1) a)

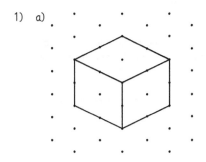

b)

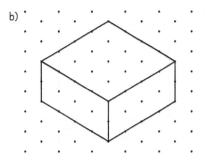

c)

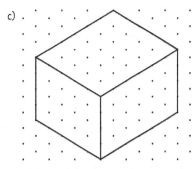

2) a)

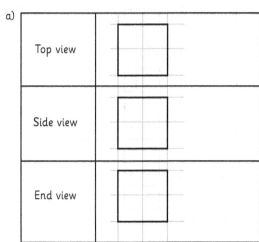

b)

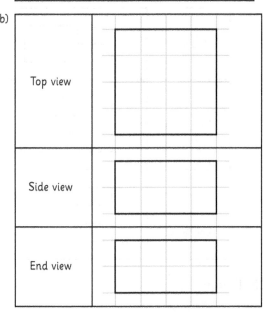

c)

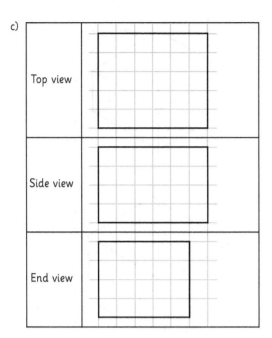

3) A

B

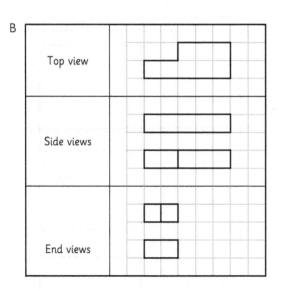

13.4 Describing and sorting prisms and pyramids (p.160)

Before we start
An appropriate justification for any of the shapes being the odd one out.

Questions
1)

3D object	A	B	C	D	E	F	G	H	I
prism	✓		✓	✓	✓			✓	✓
not a prism		✗				✗	✗		

2)

3D object	A	B	C	D	E	F	G	H
number of edges	12	8	12	9	18	10	6	15
number of vertices	8	5	8	6	12	6	4	10

3)

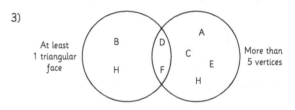

Challenge
The number of edges is equal to half the number of vertices × 3

14 Angles, symmetry and transformation

14.1 Identifying angles (p.162)

Before we start
   b) Turn 180°

Questions
1) a) acute   b) right

     c) obtuse   d) right

     e) acute   f) obtuse

2) A – right; B – obtuse; C – acute; D – acute; E – obtuse; F – right; G – obtuse; H – acute; I – obtuse; J – right

Challenge
Finlay is correct. The angle measures 95 degrees, therefore it is an obtuse angle.

14.2 Using an eight-point compass (p.164)

Before we start
Isla will be facing East.

Questions
1) a) i) Café   ii) Football stadium
      iii) Supermarket   iv) School
   b) i) Library   ii) Hotel
      iii) Hotel   iv) Bank
   c) Library, supermarket

2 a) Beach   b) Big rock
   c) Wood   d) Log pile

3)

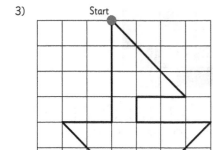

Challenge
Instructions will vary. Here is one solution (12 moves).

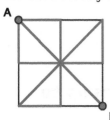

E1-S1-NW1-S1-E1-SW1-E1-N1-NE1-S1-W1-SE1

14.3 Plotting points using coordinates (p.168)

Before we start
Nuria is wrong: the triangle is at C4. The horizontal reference is always first.

Questions
1) a) The points are in a straight line.

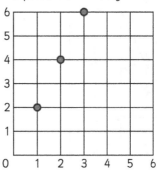

   b) The points make a triangle.

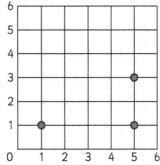

2) a)

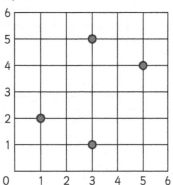

   (4, 2)

   b) (3, 1)

3) (5, 1)

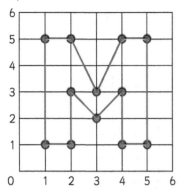

Challenge

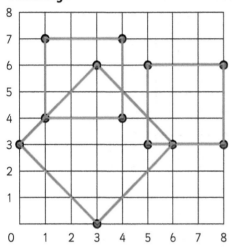

14.4 Lines of symmetry (p.170)

Before we start
An appropriate justification for any of the pictures being the odd one out.

Questions

1) A, C, F, G

2)

Horizontal line of symmetry	Vertical line of symmetry	No line of symmetry
J M O	I K P	L N

3)

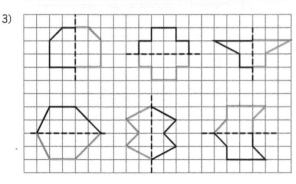

Challenge
The shape will have eight sides. The bottom line is not horizontal, so will make two sides when reflected.

14.5 Creating designs with lines of symmetry (p.173)

Before we start
Answers will vary.

Questions

1) a)

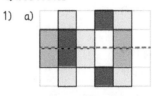

b)

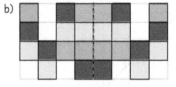

2)

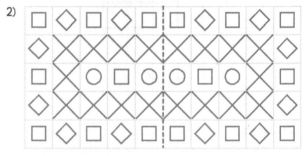

3) Check students' drawings.

Challenge
Answers will vary.

14.6 Measuring angles up to 180° (p.176)

Before we start
KIT

Questions

1) a) 30° b) 90° c) 50° d) 140° e) 120° f) 20°

2) a) 37° b) 106° c) 142° d) 53° e) 37° f) 70°
   g) 159°

Challenge
Rule: the angles in a triangle add up to 180°.

14.7 Understanding scale (p.179)

Before we start

Amman is right. There are 1000 m in 1 km, so there are 2000 m in 2 km.

Questions

1) a) 6 km   b) 9 km

2) a) 25 km   b) 75 km   c) 250 km

3) a) 400 km   b) 200 km   c) 350 km

Challenge

    a) Perth   b) 40 km   c) 120 km

    d) Edinburgh and Glasgow.

15 Data handling and analysis

15.1 Reading and interpreting information (p.182)

Before we start

There is no scale on the bar graph, so it is impossible to say how many animals there are at the aquarium.

Questions

1) a) 30 more people.

    b) 30 more people.

    c) Salt and vinegar.

    d) 145 people took part in the survey.

2) a) 40 more   b) 20 more   c) Day 2   d) 160

3) a) 300   b) 500   c) 100   d) 3400

Challenge

She should display her data in a line graph.

The first question is more easily answered by calculating the difference between Monday and Sunday, using the table. The second question is more easily answered by looking at how steep the lines are in the graph, and choosing the steepest.

15.2 Organising and displaying data (p.187)

Before we start

Finlay should label the axis with the names of the birds 'Type of bird' and the axis with numbers 'Number of birds seen'.

Questions

1)

Snack	Tally	Frequency
Crisps	IIII IIII IIII IIII IIII IIII IIII I	36
Fruit	IIII IIII IIII I	16
Sweets	IIII IIII IIII IIII IIII IIII II	32
Biscuits	IIII IIII II	12
Vegetable sticks	IIII III	8

    a) 24

    b) 44

    c) 28

    d)

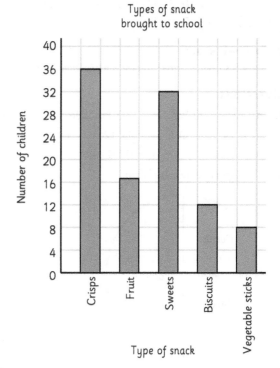

Types of snack brought to school

2)

Age in years	Tally	Number of children (frequency)
6	IIII	5
7	IIII I	6
8	IIII IIII	10
9		0
10	IIII I	6

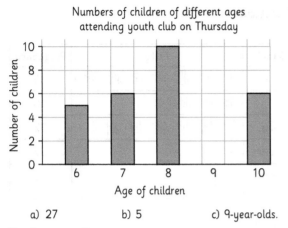

Numbers of children of different ages
attending youth club on Thursday

a) 27   b) 5   c) 9-year-olds.

3) Answers will vary.

Challenge
Variable according to the survey conducted.

15.3 Reading and interpreting pie charts (p.190)

Before we start
$32 \div 4 = 8$ so Isla is correct to give Amman eight strawberries.

Questions
1) a) 80   b) Vanilla   c) Pistachio   d) 10
2) a) i) 24   ii) 6   iii) 12   iv) 6
   b) Football   c) 18
3)   Types of vegetable planted

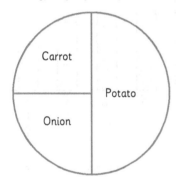

Challenge
Same: Children walk and go by car to both schools.
The same fraction of children in both schools go by car.
Different: More children walk to school A.
Some children go by bus to school B.
Children at school B travel to school in a greater variety of ways.
The pie charts might tell us that children in school A live nearer to the school than the children in school in B.

15.4 Collecting data (p.192)

Before we start
C best represents someone walking to the post box and back. The person walks increasingly far from home until they reach the postbox, spends no time there and starts heading back, walking increasing close to home. In B the person has not walked back, and in A the person has stopped for a time before turning back.

Questions
1)   Category data (words) – b, d
      Category data (numbers) – c
      Time series data – a, e, f

2)   **Bar graph (A)**
      I wonder how the number of trees differs between different areas.
      I wonder which famous places we would like to visit.
      Line graph (B)
      I wonder how many concert tickets are sold each day.

3)   a)

Number of cars	Number of people
0	
1–2	
3–4	
5–6	
7 or more	

b)

Age on 1st January	Height (cm)
10	
11	
12	
13	
14	

Challenge
a) Appropriate big question chosen, and correct form of data chosen for the inquiry.

b) The table contains suitable headings, and the bar/line graph has a suitable title and labelled axes.

16 Ideas of chance and uncertainty

16.1 Predicting and explaining simple chance situations (p.196)

Before we start

a) Certain   b) Even chance

c) Impossible   d) Likely

Questions

1) Answers will vary.

2) I will go to school tomorrow: Likely.

I will be older next week: Certain.

I like ice-cream: Answers will vary.

When I wake up it will be dark: Answers will vary.

I draw a counter from a bag…:   Even chance.

3) a) The likelihood is the same.

b) The likelihood is the same.

Challenge

1) H-H, H-T, T-T, T-H: four outcomes

2) H-H-H, H-H-T, H-T-T, T-T-T, T-T-H, T-H-H, H-T-H, T-H-T eight outcomes

3) H-H-H-H, H-H-H-T, H-H-T-H, H-T-H-H, T-H-H-H, H-H-T-T, H-T-H-T, H-T-T-H, T-H-H-T, T-H-T-H, T-T-H-H, H-T-T-T, T-H-T-T, T-T-H-T, T-T-T-H, T-T-T-T: 16 outcomes

4) 32 outcomes as it doubles each time.

© 2018 Leckie

001/05122018

11

The authors assert their moral rights to be identified as the authors for this work.

ISBN 9780008313982

Published by
Leckie
An imprint of HarperCollins*Publishers*
Westerhill Road, Bishopbriggs, Glasgow, G64 2QT
T: 0844 576 8126   F: 0844 576 8131
leckiescotland@harpercollins.co.uk   www.leckiescotland.co.uk

HarperCollins Publishers
Macken House, 39/40 Mayor Street Upper, Dublin 1, D01 C9W8, Ireland

Publisher: Fiona McGlade
Managing editor: Craig Balfour
Project editor: Rachel Allegro

Special thanks
Answer checking: Rodger Alderson
Copy editor: Gwynneth Drabble
Cover design: Ink Tank
Layout and illustration: Jouve
Proofreader: Dylan Hamilton

A CIP Catalogue record for this book is available from the British Library.

Acknowledgements
Whilst every effort has been made to trace the copyright holders, in cases where this has been unsuccessful, or if any have inadvertently been overlooked, the Publishers would gladly receive any information enabling them to rectify any error or omission at the first opportunity.

This book contains FSC™ certified paper and other controlled sources to ensure responsible forest management.

For more information visit: www.harpercollins.co.uk/green